Tibor Radó

On the Problem of Plateau

Subharmonic Functions

Reprint

Springer-Verlag Berlin Heidelberg New York 1971

Tibor Radó

On the Problem of Plateau

Subharmonic Functions

Reprint

Springer-Verlag New York Heidelberg Berlin 1971

AMS Subject Classifications (1970): 49F10, 31C05

ISBN-13:978-3-540-05479-5 e-ISBN-13:978-3-642-65236-3
DOI: 10.1007/978-3-642-65236-3

Library of Congress Catalog Card Number 71-160175.

Herstellung: Strauss & Cramer, Leutershausen

ERGEBNISSE DER MATHEMATIK
UND IHRER GRENZGEBIETE
HERAUSGEGEBEN VON DER SCHRIFTLEITUNG
DES
„ZENTRALBLATT FÜR MATHEMATIK"
ZWEITER BAND
2

ON THE PROBLEM OF PLATEAU

BY

TIBOR RADÓ

WITH 1 FIGURE

BERLIN
VERLAG VON JULIUS SPRINGER
1933

Contents.

Introduction.

The most immediate one-dimensional variation problem is certainly the problem of determining an arc of curve, bounded by two given points and having a smallest possible length. The problem of determining and investigating a surface with given boundary and with a smallest possible area might then be considered as the most immediate two-dimensional variation problem.

The classical work, concerned with the latter problem, is summed up in a beautiful and enthusiastic manner in DARBOUX's Théorie générale des surfaces, vol. I, and in the first volume of the collected papers of H. A. SCHWARZ. The purpose of the present report is to give a picture of the progress achieved in this problem during the period beginning with the Thesis of LEBESGUE (1902).

Our problem has always been considered as the outstanding example for the application of Analysis and Geometry to each other, and the recent work in the problem will certainly strengthen this opinion. It seems, in particular, that this recent work will be a source of inspiration to the Analyst interested in Calculus of Variations and to the Geometer interested in the theory of the area and in the theory of the conformal maps of general surfaces. These aspects of the subject will be especially emphasized in this report.

The report consists of six Chapters. The first three Chapters are concerned with investigations which yielded either important tools or important ideas for the proofs of the existence theorems reviewed in the last three Chapters.

Chapter I.

Curves and surfaces.

I.1. If $x = x(t)$, $y = y(t)$, $z = z(t)$, $a \leqq t \leqq b$ are the equations of a curve C, then under the usual classroom assumptions the length $l(C)$ of C is given by the formula

$$l(C) = \int_a^b \left[\left(\frac{dx}{dt}\right)^2 + \left(\frac{dy}{dt}\right)^2 + \left(\frac{dz}{dt}\right)^2\right]^{\frac{1}{2}} dt. \qquad (1.1)$$

If C reduces to a straight segment of length l, then the formula (1.1) reduces to $l = (l_1^2 + l_2^2 + l_3^2)^{\frac{1}{2}}$, where l_1, l_2, l_3 denote the lengths of the orthogonal projections of the segment upon the axes x, y, z (the coordinate system will always be supposed to be rectangular). The formula (1.1) is equally evident geometrically if C is a polygon. It is then clear that for a general curve C the formula results by approximating C by polygons[1]. As a matter of fact, (1.1) follows immediately by approximating C by an inscribed polygon.

I.2. If $x = x(u, v)$, $y = y(u, v)$, $z = z(u, v)$, (u, v) in some region R, are the equations of a surface S, then under the usual classroom assumptions the area $\mathfrak{A}(S)$ of S is given by the formula

$$\mathfrak{A}(S) = \iint_R \left[\left(\frac{\partial(y, z)}{\partial(u, v)}\right)^2 + \left(\frac{\partial(z, x)}{\partial(u, v)}\right)^2 + \left(\frac{\partial(x, y)}{\partial(u, v)}\right)^2\right]^{\frac{1}{2}} du\, dv. \qquad (1.2)$$

If S reduces to a triangle with area Δ, then (1.2) reduces to

$$\Delta = (\Delta_1^2 + \Delta_2^2 + \Delta_3^2)^{\frac{1}{2}},$$

where $\Delta_1, \Delta_2, \Delta_3$ denote the areas of the triangles obtained by orthogonal projection upon the planes yz, zx, xy. The formula (1.2) is equally evident geometrically if S is a polyhedron. It is then clear that the formula (1.2) *should* result by approximating S by polyhedrons. At any rate, this is the point of view which is significant for the problem of PLATEAU. However, the situation is much more complicated than in the case of the length.

I.3. The situation can be strikingly illustrated by the famous example of H. A. SCHWARZ[2]. Let S be the surface

$$S: x^2 + y^2 = 1, \qquad 0 \leqq z \leqq 1.$$

[1] By a *general* curve we mean here one which is *not a polygon*. For an *actually* general continuous curve (1.1) is generally wrong. Cf. I.11.

[2] Gesammelte Mathematische Abhandlungen, vol. I pp. 309–311. We have slightly changed the notations of SCHWARZ.

Cut S along the generator $x = 1$, $y = 0$, $0 \leqq z \leqq 1$, and then spread S upon a plane. The result is a rectangle R with sides 1 and 2π. Hence $\mathfrak{A}(S) = 2\pi$. Subdivide the sides of R into m and n parts respectively. Subdivide R, by parallels to the sides through the points of division, into mn congruent rectangles r. Subdivide every one of these rectangles into two triangles by drawing a diagonal. Thus R is subdivided into a network of $2mn$ triangles. Bend R so as to obtain S, and use the vertices of the network as the vertices of an inscribed polyhedron. The area $\mathfrak{A}_{m,n}$ of this polyhedron is given by

$$\mathfrak{A}_{m,n} = 2n \sin \frac{\pi}{n},$$

and hence $\mathfrak{A}_{m,n} \to 2\pi = \mathfrak{A}(S)$ for $m, n \to \infty$, which is all right. Subdivide, however, every one of the rectangles r into four triangles by drawing both diagonals. There results an inscribed polyhedron, the area $\mathfrak{A}^*_{m,n}$ of which is given by

$$\mathfrak{A}^*_{m,n} = 2n \sin \frac{\pi}{2n} + \left[\frac{1}{4} + \frac{4m^2}{n^4}\left(n \sin \frac{\pi}{2n}\right)^4\right]^{\frac{1}{2}} \times 2n \sin \frac{\pi}{n}.$$

Since we used this time a finer subdivision, it might be expected that we get a better approximation, which is however obviously not the case. Indeed, if $m = n^3$, then $\mathfrak{A}^*_{m,n} \to \infty$. If $m = n$, then $\mathfrak{A}^*_{m,n} \to 2\pi = \mathfrak{A}(S)$. Since always

$$\mathfrak{A}^*_{m,n} \geqq 2n \sin \frac{\pi}{2n} + n \sin \frac{\pi}{n},$$

it is clear that if $m, n \to \infty$ in any manner, then $\mathfrak{A}^*_{m,n}$ never converges to a limit $< 2\pi$. On the other hand it is obvious that every number k such that $2\pi \leqq k \leqq +\infty$ can be obtained as the limit of $\mathfrak{A}^*_{m,n}$, if m, n both go to infinity in a proper way.

Hence the area of inscribed polyhedrons, approximating a given surface S, do not converge, in general, to $\mathfrak{A}(S)$. This fact invalidates the geometrical interpretation of the formula (1.2) which was generally accepted before the example of SCHWARZ became known. A great number of new interpretations of (1.2) have since been proposed. In most cases, the idea of approximating the given surface by polyhedrons has been altogether dropped. However, as far as the problem of PLATEAU is concerned, the most essential facts concerning the area have been brought to light in efforts to clear up the relation between the area of a surface and the areas of approximating polyhedrons. We are going to give a brief account of the theory of the area from this point of view.

I.4. The first thing is to define the area of a surface. Of the many definitions which have been proposed only the definition given by LEBESGUE in his Thesis[1] became significant for the

[1] Intégrale, longueur, aire. Ann. Mat. pura appl. Vol. 7 (1902) pp. 231–359.

problem of PLATEAU, and therefore only that definition will be considered here[1].

In the example of SCHWARZ (see I.3) the areas of the approximating polyhedrons showed the tendency of converging to values larger than the area of the given surface S. The definition of the area given by LEBESGUE is based on the intuitive assumption that this tendency is absolutely general: if a sequence of surfaces converges to a surface, then the areas never converge to a value less than the area of the limit surface. Given then a class of surfaces S, we wish to define the area $\mathfrak{A}(S)$ of S in such a way that the above intuitive assumption be satisfied, that is to say in such a way that

$$\underline{\lim}\,\mathfrak{A}(S_n) \geqq \mathfrak{A}(S) \quad \text{if} \quad S_n \to S.$$

In other words, $\mathfrak{A}(S)$ has to be a lower semi-continuous functional. We also require that it must be possible to compute $\mathfrak{A}(S)$ by approximating S by polyhedrons; in other words, we require that there exists, for every surface S, a sequence of polyhedrons $\mathfrak{P}_n$ such that

$$\mathfrak{P}_n \to S \quad \text{and} \quad \mathfrak{A}(\mathfrak{P}_n) \to \mathfrak{A}(S).$$

Finally, we require that if S is a polyhedron, then $\mathfrak{A}(S)$ is equal to the area of the polyhedron in the elementary sense. These three conditions determine $\mathfrak{A}(S)$ univocally. Indeed, for every sequence of polyhedrons $\mathfrak{P}_n$ converging to S we must have $\underline{\lim}\,\mathfrak{A}(\mathfrak{P}_n) \geqq \mathfrak{A}(S)$, while the sign of equality holds for at least one sequence $\mathfrak{P}_n$. That is to say, $\mathfrak{A}(S)$ is the smallest value which is the limit of the areas of polyhedrons converging to S. This is the definition of the area given by LEBESGUE.

This definition, if it is to be consistent, implies the theorem that if a sequence of polyhedrons $\mathfrak{P}_n$ converges to a polyhedron $\mathfrak{P}$, then $\underline{\lim}\,\mathfrak{A}(\mathfrak{P}_n) \geqq \mathfrak{A}(\mathfrak{P})$, where $\mathfrak{A}$ denotes the area in the elementary sense (that is to say the sum of the areas of the faces of the polyhedron). Besides, the notions used in the definition must first be clearly defined. These points will be considered later on. For the moment, we wish to call a few peculiar facts to the attention of the reader.

I.5. Suppose S consists of the points in and on a JORDAN curve C situated in a plane. As is well known, the two-dimensional measure of C might be positive, and therefore the question arises as to whether $\mathfrak{A}(S)$ is the interior or the exterior measure. Since C can be approximated by polygons from the inside, it follows readily that $\mathfrak{A}(S)$ is at most equal to the interior area, that is to say to the measure of the open domain bounded by C.

Now, one of the most natural assumptions concerning the area is this: if a surface is projected orthogonally upon a plane, then the area

[1] For literature and a systematic presentation, see T. RADÓ: Über das Flächenmaß rektifizierbarer Flächen. Math. Ann. Vol. 100 (1928) pp. 445–479.

of the surface is at least equal to the measure of the projection. In the above example, the projection is the closed region bounded by C. Hence, if the two-dimensional measure of C is positive, we have an example showing that *the area of a surface is in general less than the measure of the orthogonal projection of the surface upon a plane.*

I.6. This situation, which is an inevitable consequence of the requirement that $\mathfrak{A}(S)$ be a lower semi-continuous functional, constitutes one of the main difficulties in handling the definition of LEBESGUE. GEÖCZE devised the following simple example which shows the situation possibly at its worst[1]. Let the surface S be given by equations

$$S: x = x(u, v), \quad y = y(u, v), \quad z = z(u, v), \quad 0 \leqq u \leqq 1, \quad 0 \leqq v \leqq 1,$$

where $x(u, v), y(u, v), z(u, v)$ are continuous. Subdivide the square $0 \leqq u \leqq 1$, $0 \leqq v \leqq 1$ into n^2 congruent squares, and subdivide every one of these smaller squares into two triangles by drawing a diagonal. Use the points of S which correspond to the vertices of this triangular net as the vertices of an inscribed polyhedron $\mathfrak{P}_n$. Then, by definition,

$$\mathfrak{A}(S) \leqq \underline{\lim}\, \mathfrak{A}(\mathfrak{P}_n) \quad \text{for} \quad n \to \infty.$$

Suppose now that $x(u, v), y(u, v), z(u, v)$ are functions of u alone:

$$x(u, v) \equiv f_1(u), \quad y(u, v) \equiv f_2(u), \quad z(u, v) \equiv f_3(u).$$

Obviously, $\mathfrak{A}(\mathfrak{P}_n) = 0$ for every n, and hence $\mathfrak{A}(S) = 0$, which looks all right, since S reduces in reality to the curve

$$\Gamma: x = f_1(u), \quad y = f_2(u), \quad z = f_3(u).$$

If we choose however Γ as a PEANO curve filling a cube, then we obtain an example showing that *a surface might contain every point of a cube and might still have a zero area.*

I.7. The definition of LEBESGUE implies a previous definition of convergent sequences of surfaces. This latter definition will be based on the notion of the distance of two surfaces. We shall now show in a simple example how important the definition of the distance is. Take two surfaces S_1, S_2. Define the distance of S_1 and S_2 as the smallest number δ with the properties: 1. for every point P_1 of S_1 there exists a point P_2 of S_2 such that the distance P_1P_2 is less than or equal to δ, and 2. for every point P_2 of S_2 there exists a point P_1 of S_1 such that the distance P_1P_2 is less than or equal to δ. Given then a sequence of surfaces S_n and a surface S, $S_n \to S$ means that the distance of S_1 and S_2 converges to zero.

Suppose we use this definition of convergence in the definition of the area (which we shall not do). Given then a continuous surface S, and an $\varepsilon > 0$, it is clear that we can take a very long and very narrow

---

[1] Z. DE GEÖCZE: Sur l'exemple d'une surface dont l'aire est égale à zéro et qui remplit un cube. Bull. Soc. Math. France (1913) pp. 29–31.

ribbon of paper, of rectangular shape and with an area less than ε, and deform the ribbon so as to obtain a polyhedron the distance of which from S, in the sense defined above, is less than ε. Doing this for $\varepsilon = 1, \frac{1}{2}, \frac{1}{3}, \ldots$, we obtain a sequence of polyhedrons $\mathfrak{P}_n$ such that $\mathfrak{A}(\mathfrak{P}_n) < \frac{1}{n}$ and $\mathfrak{P}_n \to S$. Consequently, $\mathfrak{A}(S) = 0$. That is to say, if we use the definition of the distance given above (which we shall not do), then the area of every continuous surface is zero.

This shows clearly that *the definition of convergent sequences of surfaces is of the greatest importance for the definition of the area* given by LEBESGUE.

I.8. We now shall give the exact definition of $\mathfrak{A}(S)$ for the class of the *continuous surfaces of the type of the circular disc.* Such a surface is defined by a set of equations

$$S: x = x(u, v), \quad y = y(u, v), \quad z = z(u, v), \quad (u, v) \text{ in } R,$$

where R denotes some JORDAN region (that is to say the set of points in and on a JORDAN curve), and $x(u, v), y(u, v), z(u, v)$ are continuous in R. We do not suppose that distinct points (u, v) are carried into distinct points (x, y, z). Given then two such surfaces

$$S_1: x = x_1(u, v), \quad y = y_1(u, v), \quad z = z_1(u, v), \quad (u, v) \text{ in } R_1, \tag{1.3}$$

and

$$S_2: x = x_2(u, v), \quad y = y_2(u, v), \quad z = z_2(u, v), \quad (u, v) \text{ in } R_2, \tag{1.4}$$

consider a topological correspondence T between R_1 and R_2. Denote by $(u_1, v_1), (u_2, v_2)$ a couple of corresponding points, and denote by P_1, P_2 the points which correspond to $(u_1, v_1), (u_2, v_2)$ by means of the equations (1.3), (1.4) respectively. The maximum of the distance P_1P_2, for all possible positions of the points $(u_1, v_1), (u_2, v_2)$ corresponding to each other under T, will be denoted by $\delta(T)$. The greatest lower bound of $\delta(T)$, for all possible topological correspondences between R_1 and R_2, is *the distance, in the sense of* FRÉCHET, of S_1 and S_2 and will be denoted by $d(S_1, S_2)$[1].

If $d(S_1, S_2) = 0$, the surfaces S_1, S_2 will be considered as identical. We shall say also, if $d(S_1, S_2) = 0$, that (1.3) and (1.4) are *parametric representations of the same surface.*

If we have a surface S and a sequence of surfaces S_n, such that $d(S_n, S) \to 0$, then we shall say that $S_n \to S$.

It should be observed that the definition of the distance, in the sense of FRÉCHET, presupposes that we are dealing with surfaces of the same topological type. In this report, except for parts of Chapter VI, we shall consider continuous surfaces of the type of the circular disc only. The term *surface* will be used in this sense, unless the contrary is explicitly stated.

[1] Sur la distance de deux surfaces. Ann. Soc. Polon. math. Vol. 3 (1924) pp. 4—19.

It is also important to remark that two surfaces S_1, S_2 might consist of the same points (x, y, z) without being identical in the sense $d(S_1, S_2) = 0$.

I.9. A surface S will be called a *polyhedron* and will be denoted by $\mathfrak{P}$ if it admits of a parametric representation

$$\mathfrak{P}: x = x(u, v), \quad y = y(u, v), \quad z = z(u, v), \quad (u, v) \text{ in } R, \qquad (1.5)$$

with the following properties. The JORDAN region R can be subdivided into a finite number of non-overlapping curvilinear triangles $\delta_1, \delta_2, \ldots, \delta_n$ in such a way that every one of these triangles is carried, by the equations (1.5), in a one-to-one and continuous way into a (non-degenerate) plane rectilinear triangle in the xyz-space. The boundary of R is carried in a one-to-one and continuous way into a simple closed polygon. Such a representation will be called a *typical representation* of $\mathfrak{P}$.

If $\Delta_1, \Delta_2, \ldots, \Delta_n$ are the (plane and rectilinear) triangles into which $\delta_1, \delta_2, \ldots, \delta_n$ are carried by the equations (1.5), then the sum of the areas of $\Delta_1, \Delta_2, \ldots, \Delta_n$ will be called the elementary area of $\mathfrak{P}$ and will be denoted by $E(\mathfrak{P})$. Then $E(\mathfrak{P})$ can be shown to be independent of the choice of the typical representation which has been used for the computation.

I.10. Given a surface S, consider a sequence $\mathfrak{P}_n$ of polyhedrons converging, in the FRÉCHET sense (see I.8) to S. Consider $\underline{\lim} E(\mathfrak{P}_n)$. Then the greatest lower bound of $\underline{\lim} E(\mathfrak{P}_n)$, for all possible sequences $\mathfrak{P}_n \to S$, is by definition the area (in the sense of LEBESGUE) of S, and will be denoted by $\mathfrak{A}(S)$. Thus $\mathfrak{A}(S)$ is defined for every continuous surface of the type of the circular disc. $\mathfrak{A}(S)$ might be $+\infty$.

This definition of $\mathfrak{A}(S)$ is consistent as it stands, and would be logically consistent even if it would not be true that if S is a polyhedron $\mathfrak{P}$, then $\mathfrak{A}(\mathfrak{P}) = E(\mathfrak{P})$. On the other hand it is clear that $\mathfrak{A}(\mathfrak{P}) = E(\mathfrak{P})$ must be true if the definition is to serve any useful purpose. The theorem $\mathfrak{A}(\mathfrak{P}) = E(\mathfrak{P})$ is true[1], but not obvious; its truth, and in a general way the usefulness of $\mathfrak{A}(S)$, depends essentially upon the definition of the distance of two surfaces, which we have decided to use (see I.7).

I.11. In developing the theory of the area $\mathfrak{A}(S)$, a good working program is obtained by setting up the principle that the theory of the area should be analogous to the theory of the length. Let us recall therefore a few facts concerning the length. If

$$C: x = x(t), \quad y = y(t), \quad z = z(t), \quad a \leqq t \leqq b$$

are the equations of a continuous curve, then the length $l(C)$ is defined as the least upper bound of the lengths of inscribed polygons. While this definition is not analogous to the definition of $\mathfrak{A}(S)$, it can easily

[1] For a direct proof, see M. FRÉCHET: Sur la semi-continuité en géométrie élémentaire, Nouvelles Ann. de Math. Vol. 3 (1924).

be shown that it is equivalent to the following definition. Define first the distance, in the FRÉCHET sense, of two continuous curves

$$C_1 : x = x_1(t), \quad y = y_1(t), \quad z = z_1(t), \quad a_1 \leqq t \leqq b_1,$$
$$C_2 : x = x_2(t), \quad y = y_2(t), \quad z = z_2(t), \quad a_2 \leqq t \leqq b_2$$

by using one-to-one and continuous correspondences between the intervals $a_1 \leqq t \leqq b_1$ and $a_2 \leqq t \leqq b_2$ in exactly the same manner as the distance of two continuous surfaces has been defined in I.8. Then the length $l(C)$ of a continuous curve C can be defined as the smallest number which is the limit of the lengths of polygons converging, in the FRÉCHET sense, to C.

On account of this definition, $l(C)$ clearly is a lower semi-continuous functional. That is to say, if $C_n \to C$, then $\underline{\lim}\, l(C_n) \geqq l(C)$. For a polygon $\mathfrak{p}$, $l(\mathfrak{p})$ is equal to the length of $\mathfrak{p}$ in the elementary sense. If a sequence $\mathfrak{p}_n$ of polygons converges to a continuous curve C, then $\underline{\lim}\, l(\mathfrak{p}_n) \geqq l(C)$. If $\mathfrak{p}_n \to C$, and $\mathfrak{p}_n$ is inscribed in C, then $\lim l(\mathfrak{p}_n) = l(C)$. This gives a convergent process for calculating $l(C)$. Let

$$C : x = x(t), \quad y = y(t), \quad z = z(t), \; a \leqq t \leqq b$$

be the equations of C. Subdivide the interval $a \leqq t \leqq b$ by points

$$a = t_0 < t_1 < \cdots < t_{k-1} < t_k < \cdots < t_n = b$$

into subintervals. Denote by D this subdivision and by $\|D\|$ the greatest length of any of the subintervals. Then

$$\lim \sum_{k=1}^{n} \{[x(t_k) - x(t_{k-1})]^2 + [y(t_k) - y(t_{k-1})]^2 + [z(t_k) - z(t_{k-1})]^2\}^{\frac{1}{2}} = l(C) \quad (1.6)$$

if $\|D\| \to 0$.

The relation between $l(C)$ and the classical integral formula is also completely known[1]. If $l(C)$ is finite, then the coordinate functions $x(t), y(t), z(t)$ are of bounded variation. The differential coefficients $x'(t)$, $y'(t)$, $z'(t)$ exist almost everywhere, and

$$\int_a^b (x'^2 + y'^2 + z'^2)^{\frac{1}{2}} dt \leqq l(C).$$

The sign of equality holds if and only if $x(t), y(t), z(t)$ are absolutely continuous. There always exist representations of C which satisfy this condition [provided $l(C)$ is finite]. For instance, if $s(t)$ denotes the length of the arc corresponding to the interval from a to t, then x, y, z as functions of $s = s(t)$ satisfy even the LIPSCHITZ condition.

I.12. With the preceding facts as a basis of comparison, the theory of the area $\mathfrak{A}(S)$ appears as being surprisingly incomplete. $\mathfrak{A}(S)$ easily is found to be a lower semi-continuous functional, and we also mentioned

[1] Cf. L. TONELLI: Sopra alcuni polinomi approssimativi. Ann. Mat. pura appl. Vol. 25 (1916).

that for polyhedrons $\mathfrak{P}$ the area $\mathfrak{A}(\mathfrak{P})$ is equal to the area in the elementary sense. If a sequence of polyhedrons $\mathfrak{P}_n$ converges to a surface S, then $\underline{\lim}\,\mathfrak{A}(\mathfrak{P}_n) \geqq \mathfrak{A}(S)$. But no general procedure is known which would permit us to construct, for every continuous surface S of the type of the circular disc, a sequence of polyhedrons $\mathfrak{P}_n$ such that $\mathfrak{P}_n \to S$ and $\mathfrak{A}(\mathfrak{P}_n) \to \mathfrak{A}(S)$, although such a sequence certainly exists by definition. It is not known if it always is possible to select a sequence $\mathfrak{P}_n$ of *inscribed* polyhedrons such that $\mathfrak{P}_n \to S$ and $\mathfrak{A}(\mathfrak{P}_n) \to \mathfrak{A}(S)$. No general convergent process is known which would permit us to compute $\mathfrak{A}(S)$ in terms of the coordinate functions $x(u, v)$, $y(u, v)$, $z(u, v)$, while formula (1.6) in I.11 gives such a process for the computation of the length of every continuous curve. It is not known if for every surface S with a finite $\mathfrak{A}(S)$ there exists a representation such that the integral (1.2) exists and is equal to $\mathfrak{A}(S)$.

On the other hand, very satisfactory results have been obtained for certain special classes of surfaces which now will be considered briefly.

I.13. Suppose S is given by an equation

$$S: z = z(x, y), \qquad (x, y) \text{ in } R,$$

where $z(x, y)$ is single-valued and continuous in the closed region R which we shall suppose to be a rectangle

$$R: a \leqq x \leqq b, \quad c \leqq y \leqq d.$$

The following expressions, introduced by the Hungarian mathematician GEÖCZE[1], are fundamental in the theory of $\mathfrak{A}(S)$. Let

$$r: x' \leqq x \leqq x'', \quad y' \leqq y \leqq y''$$

denote a rectangle r comprised in R. Put:

$$\alpha(z, r) = \int_{x'}^{x''} |z(\xi, y'') - z(\xi, y')|\, d\xi,$$

$$\beta(z, r) = \int_{y'}^{y''} |z(x'', \eta) - z(x', \eta)|\, d\eta,$$

$$\gamma(r) = (x'' - x')(y'' - y'),$$

$$g(z, r) = (\alpha^2 + \beta^2 + \gamma^2)^{\frac{1}{2}}.$$

Consider then a subdivision D of R into rectangles, obtained by drawing a finite number of parallels to the x and y axes respectively. Denote by $\|D\|$ the greatest diagonal of the rectangles of D. Put

$$G(z, D) = \sum_{(r)} g(z, r),$$

[1] Quadrature des surfaces courbes. Math. naturwiss. Ber., Ungarn Vol. 26 (1910) pp. 1–88.

where the summation is extended over all the rectangles r of D. Then $G(z, D)$ will be called a GEÖCZE sum for the surface S.

GEÖCZE proved the following facts concerning these sums. If D^* is a subdivision of D, then $G(z, D) \leqq G(z, D^*)$. If $z_n(x, y) \rightrightarrows z(x, y)$ in R, then $G(z_n, D) \to G(z, D)$ for every fixed subdivision D. For every subdivision D we have the inequality $G(z, D) \leqq \mathfrak{A}(S)$. If D_n is a sequence of subdivisions such that $\|D_n\| \to 0$, then $G(z, D_n)$ converges to a definite limit independent of the particular choice of the sequence D_n. Put

$$\Gamma(z) = \lim G(z, D) \quad \text{for} \quad \|D\| \to 0 .$$

$\mathfrak{A}(S)$ is finite if and only if $\Gamma(z)$ is finite.

GEÖCZE also stated the theorem $\Gamma(z) = \mathfrak{A}(z)$. He proved it however only in the special case when $z(x, y)$ satisfies the LIPSCHITZ condition. In this case he proved also that

$$G(z, D) \to \iint_R (1 + p^2 + q^2)^{\frac{1}{2}}\, dx\, dy \quad \text{for} \quad \|D\| \to 0,$$

where the partial derivatives $p = z_x$, $q = z_y$ exist almost everywhere and are bounded and measurable, on account of the LIPSCHITZ condition. Thus GEÖCZE proved that if $z(x, y)$ satisfies the LIPSCHITZ condition, then $\mathfrak{A}(S)$ is given by the classical formula

$$\mathfrak{A}(S) = \iint_R (1 + p^2 + q^2)^{\frac{1}{2}}\, dx\, dy .$$

Suppose now only that $\mathfrak{A}(S)$ is finite. Then TONELLI[1] proved that the partial derivatives $p = z_x$ and $q = z_y$ exist almost everywhere in R, that $(1 + p^2 + q^2)^{\frac{1}{2}}$ is summable, and that

$$\iint_R (1 + p^2 + q^2)^{\frac{1}{2}}\, dx\, dy \leqq \mathfrak{A}(S) .$$

TONELLI obtained then the beautiful theorem that the sign of equality holds if and only if $z(x, y)$ is absolutely continuous[2]. By a combination of the methods of GEÖCZE and TONELLI, RADÓ[3] proved the theorem, stated without proof by GEÖCZE, that

$$G(z, D) \to \mathfrak{A}(S) \quad \text{for} \quad \|D\| \to 0, \tag{1.6.1}$$

under the assumption only that $z(x, y)$ is continuous. The theorem holds even if $\mathfrak{A}(S)$ is infinite. (1.6.1) gives a convergent process for

[1] Sulla quadratura delle superficie. Rend. Accad. naz. Lincei Vol. 3 (1926) pp. 445—450.

[2] L. TONELLI: Sulla quadratura delle superficie. Rend. Accad. naz. Lincei Vol. 3 (1926) pp. 633—638.

[3] Sur le calcul de l'aire des surfaces courbes. Fundam. Math. Vol. 10 (1927) pp. 197—210. A similar result has been obtained later by TONELLI: Sulla quadratura delle superficie. Rend. Accad. naz. Lincei Vol. 5 (1927) pp. 313—318.

the area $\mathfrak{A}(S)$ of every continuous surface S of the form $z = z(x, y)$, and this process, as far as the appearance of the formulas is concerned, is absolutely analogous to the process given by (1.6) in I.11 for the length of a curve. It should be observed however that $G(z, D)$ cannot be interpreted as the area of a polyhedron. The problem of constructing a sequence $\mathfrak{P}_n$ of polyhedrons such that $\mathfrak{P}_n \to S$ and $\mathfrak{A}(\mathfrak{P}_n) \to \mathfrak{A}(S)$ is still unsolved even if stated only for continuous surfaces of the special form $z = z(x, y)$.

Since (1.6.1) gives an analytic expression for $\mathfrak{A}(S)$, it might be expected that the theory of $\mathfrak{A}(S)$ can be based solely on that analytic expression and is thus accessible to general analytic methods, without further reference to the geometrical definition of $\mathfrak{A}(S)$. This program has been stated and carried out by S. SAKS[1]. The basic theorem in his work is as follows. Let (x_0, y_0) be an interior point of the rectangle R, and denote by σ a small square comprising (x_0, y_0). Denote by S_σ the portion of S situated above σ. Then, if the diameter of σ converges to zero,

$$\frac{\mathfrak{A}(S_\sigma)}{\text{area of } \sigma} \to [1 + p(x_0, y_0)^2 + q(x_0, y_0)^2]^{\frac{1}{2}}$$

for almost every point (x_0, y_0) in R (that is to say for every point in R with the possible exception of a set of measure zero). It is presupposed in the theorem that $\mathfrak{A}(S)$ is finite.

I.14. For surfaces of the general type

$$S: x = x(u, v), \quad y = y(u, v), \quad z = z(u, v), \quad (u, v) \text{ in } R, \qquad (1.6.2)$$

the theory of $\mathfrak{A}(S)$ is very incomplete. The first significant result is again due to GEÖCZE, and is concerned with the so-called rectifiable surfaces. This very misleading term is used to denote surfaces S which admit of a representation (1.6.2) such that $x(u, v)$, $y(u, v)$, $z(u, v)$ satisfy the LIPSCHITZ condition in R. Such a representation will be called a *typical representation of the rectifiable surface S*. GEÖCZE proved[2] that for every rectifiable surface S, given in typical representation, we have

$$\mathfrak{A}(S) = \iint_R \left[\left(\frac{\partial(y, z)}{\partial(u, v)}\right)^2 + \left(\frac{\partial(z, x)}{\partial(u, v)}\right)^2 + \left(\frac{\partial(x, y)}{\partial(u, v)}\right)^2\right]^{\frac{1}{2}} du\, dv.$$

The rectifiable surfaces have also been studied by RADEMACHER[3]. Suppose, for the sake of simplicity, that R is the square

$$R: 0 \leqq u \leqq 1, \quad 0 \leqq v \leqq 1.$$

[1] Sur l'aire des surfaces $z = f(x, y)$. Acta Litt. Sci. Szeged Vol. 3 (1927) pp. 170—176.

[2] This part of the work of GEÖCZE has only been published in Hungarian. For references, and for a simplified presentation of the results of GEÖCZE, see T. RADÓ: Über das Flächenmaß rektifizierbarer Flächen. Math. Ann. Vol. 100 (1928) pp. 445—479.

[3] Über partielle und totale Differenzierbarkeit I, II. Math. Ann. Vol. 79 (1919) pp. 340—359 and Vol. 81 (1920) pp. 52—63.

Subdivide R into a finite number of non-overlapping (rectilinear) triangles. Denote by $\mathfrak{N}$ this triangular net, by $\|\mathfrak{N}\|$ the greatest side of the triangles of $\mathfrak{N}$, by φ the smallest angle occurring in the triangles of $\mathfrak{N}$, and by $\mathfrak{P}$ the polyhedron whose vertices are the points of the surface S corresponding to the vertices of $\mathfrak{N}$ by means of the equations (1.6.2) of the rectifiable surface S given in typical representation. Then RADEMACHER proves that $\mathfrak{A}(\mathfrak{P})$ converges to the classical integral if $\|\mathfrak{N}\| \to 0$, provided φ remains larger than a fixed positive number. A more general criterion is obtained in terms of the greatest angle Φ occurring in the triangles of $\mathfrak{N}$, namely the criterion that $\mathfrak{A}(\mathfrak{P})$ converges to the classical integral for $\|\mathfrak{N}\| \to 0$, provided $\pi - \Phi$ remains larger than a fixed positive number. This criterion has been stated, for the elementary case when $x(u,v)$, $y(u,v)$, $z(u,v)$ have continuous derivatives, by FRÉCHET[1]. The methods of RADEMACHER permit us to generalize the criterion to rectifiable surfaces given in typical representation. Thus the problem of determining a sequence of polyhedrons $\mathfrak{P}_n$ such that $\mathfrak{P}_n \to S$ and $\mathfrak{A}(\mathfrak{P}_n) \to \mathfrak{A}(S)$ is solved for rectifiable surfaces S. The method of RADEMACHER is based on the fact that a rectifiable surface S has a definite tangent plane almost everywhere and that the restrictions upon the net $\mathfrak{N}$ imply that the faces of the corresponding inscribed polyhedron $\mathfrak{P}$ approximate the tangent planes of S.

I.15. This situation and the surmised analogy with the theory of the length might suggest a generalization whose impossibility however is demonstrated by a curious remark of S. SAKS to the effect that *a surface S with finite area $\mathfrak{A}(S)$ need not have a tangent plane anywhere*[2], not even if the surface admits a representation of the very special type $z = z(x, y)$. The reason for this might be explained as follows. Let the surface S coincide with the square $0 \leqq x \leqq 1, 0 \leqq y \leqq 1$, $z \equiv 0$. Take an interior point (x_0, y_0), a small $\varrho > 0$, and replace the circular disc $(x - x_0)^2 + (y - y_0)^2 \leqq \varrho^2$ by a cone of revolution of height h. No matter how large h is, the area of this cone will be arbitrary small if ϱ is sufficiently small. Thus we obtain from S a new surface the area of which is very slightly larger than the area of S, while the smoothness of S has been badly disturbed around (x_0, y_0). The example of S. SAKS is then obtained by putting on sharper and sharper cones and still keeping the area finite. The process is similar to that which gives continuous curves without tangents[3], with the essential difference that in the case of a curve a disturbance of the smoothness appreciably

[1] Note on the area of a surface. Proc. London Math. Soc. Vol. 24 (1926).

[2] The following is based on an oral communication of S. SAKS. — (*Added in proof.*) For the details, see his paper, On the surfaces without tangent planes, Ann. of Math. Vol. 34 (1933) pp. 114—124.

[3] See for instance K. KNOPP: Ein einfaches Verfahren etc. Math. Z. Vol. 2 (1918) pp. 1—26.

increases the length and therefore the total disturbance of the smoothness drives the length to infinity.

The long thin cones with small areas are responsible for another serious evil which will be discussed in III.13.

I.16. Important contributions to the theory of the area are due to recent investigations of E. J. McShane[1] and of C. B. Morrey[2]. We restrict ourselves to state just one theorem which will be needed later. Let there be given a continuous surface

$$S: x = x(u, v), \quad y = y(u, v), \quad z = z(u, v), \quad u^2 + v^2 \leqq 1.$$

Suppose that

1. $x(u, v)$, $y(u, v)$, $z(u, v)$ are absolutely continuous functions of u for almost every v and absolutely continuous functions of v for almost every u and

2. the Dirichlet integrals

$$\iint (x_u^2 + x_v^2), \quad \iint (y_u^2 + y_v^2), \quad \iint (z_u^2 + z_v^2),$$

taken over $u^2 + v^2 < 1$, are finite.

Then the area of S is given by the classical integral formula (1.2).

I.17. We shall also need, in Chapter VI, a result of McShane[3] concerning the so-called inequality of Steiner. Denote by R a Jordan region in the xy-plane, and consider two continuous surfaces

$$S_1: z = z_1(x, y), \quad S_2: z = z_2(x, y), \quad (x, y) \text{ in } R.$$

Denote by S the surface

$$S: z = \tfrac{1}{2}(z_1 + z_2), \quad (x, y) \text{ in } R.$$

Then

$$\mathfrak{A}(S) \leqq \tfrac{1}{2}(\mathfrak{A}(S_1) + \mathfrak{A}(S_2)). \tag{1.7}$$

This is the inequality of Steiner. It is obvious for polyhedrons, and a passage to the limit proves it for the general case. It is important for the sequel to discuss the sign of equality. Let us restrict ourselves to the case when R is a rectangle with sides parallel to the axes. Subdivide R into smaller rectangles r of the same type. For every r we have then the inequality

$$\mathfrak{A}(S^{(r)}) \leqq \tfrac{1}{2}(\mathfrak{A}(S_1^{(r)}) + \mathfrak{A}(S_2^{(r)})), \tag{1.8}$$

where $S^{(r)}$, $S_1^{(r)}$, $S_2^{(r)}$ denote the portions of S, S_1, S_2 corresponding to r. Adding up these inequalities, it follows that the sign of equality in

[1] Integrals over surfaces in parametric form. Bull. Amer. Math. Soc. Vol. 38 (1932) p. 810.

[2] A class of representations of manifolds. Bull. Amer. Math. Soc. Vol. 38 (1932) p. 809. An analytic criterion that a surface possess finite Lebesgue area. Bull. Amer. Math. Soc. Vol. 39 (1933) p. 41.

[3] On a certain inequality of Steiner: Ann. of Math. Vol. 33 (1932) pp. 123—138.

(1.7) holds if and only if it holds in (1.8) for every rectangle r comprised in R, or, in other words, if and only if the function of rectangles

$$F(r) = \tfrac{1}{2}\left(\mathfrak{A}(S_1^{(r)}) + \mathfrak{A}(S_2^{(r)})\right) - \mathfrak{A}(S^{(r)})$$

vanishes identically. Take then any point (x, y) interior to R and let r be any small square, with sides parallel to the axes, comprising (x, y). Denote by $a(r)$ the area of r. Then, on account of the theorem of S. Saks, stated at the end of I.13,

$$\frac{F(r)}{a(r)} \to \frac{1}{2}\left((1 + p_1^2 + q_1^2)^{\frac{1}{2}} + (1 + p_2^2 + q_2^2)^{\frac{1}{2}}\right) - (1 + p^2 + q^2)^{\frac{1}{2}} \qquad (1.9)$$

for almost every (x, y) in R. As $F(r) \equiv 0$, it follows that the right-hand side of (1.9) vanishes almost everywhere in R. As

$$p = \tfrac{1}{2}(p_1 + p_2), \quad q = \tfrac{1}{2}(q_1 + q_2),$$

this implies that $p_1 = p_2$, $q_1 = q_2$ almost everywhere in R. That is to say: *if the sign of equality holds in* (1.7), *then* $p_1 - p_2 = 0$, $q_1 - q_2 = 0$ *almost everywhere in* R.

From this it does *not* follow, in general, that $z_2 - z_1$ is constant in R. An important case in which this conclusion is legitimate is the case when $\mathfrak{A}(S_1)$ and $\mathfrak{A}(S_2)$ are both given by the classical integral formulas

$$\mathfrak{A}(S_1) = \iint (1 + p_1^2 + q_1^2)^{\frac{1}{2}},$$

$$\mathfrak{A}(S_2) = \iint (1 + p_2^2 + q_2^2)^{\frac{1}{2}}.$$

Indeed, according to Tonelli (see I.13), z_2 and z_1 are then both absolutely continuous, and so is therefore $z_2 - z_1$. The vanishing of $p_2 - p_1$, $q_2 - q_1$ almost everywhere in R actually implies then that $z_2 - z_1$ is constant.

I.18. In the theory of minimal surfaces, conformal mapping plays an important part. We are going to recall a few significant facts concerned with such maps. Consider a surface

$$S: x = x(u, v), \quad y = y(u, v), \quad z = z(u, v), \quad u^2 + v^2 \leqq 1. \qquad (1.10)$$

Take an interior point (u_0, v_0), and map a vicinity V_0 of (u_0, v_0) upon a vicinity V_0^* of some point (α_0, β_0) in the $\alpha\beta$-plane in a one-to-one and continuous way by equations $\alpha = \alpha(u, v)$, $\beta = \beta(u, v)$. Suppose that x, y, z, as functions of α, β, have continuous partial derivatives of the first order in V_0^* which satisfy there the relations $E^* = G^*$, $F^* = 0$, where

$$E^* = x_\alpha^2 + y_\alpha^2 + z_\alpha^2,$$
$$F^* = x_\alpha x_\beta + y_\alpha y_\beta + z_\alpha z_\beta,$$
$$G^* = x_\beta^2 + y_\beta^2 + z_\beta^2.$$

Then we shall say that α, β are *local isothermic parameters*. If $E^* = G^* > 0$ in V_0^*, then the correspondence between V_0^* and S is one-to-one and conformal, that is to say angles are preserved under the correspondence.

In the sequel, we shall have to permit the vanishing of $E^* = G^*$, that is to say we shall use the term isothermic in a more general sense than is usual in differential geometry.

The most general condition, known at present, for the existence of local isothermic parameters is as follows[1]. The surface S, given by (1.10), admits of local isothermic parameters if the following conditions are satisfied.

1. $x(u, v)$, $y(u, v)$, $z(u, v)$ have continuous partial derivatives of the first order.

2. The partial derivatives of the first order satisfy, in every closed region R entirely interior to the unit circle, the LIPSCHITZ-HÖLDER condition with some exponent λ *, possibly depending upon R, such that $0 < \lambda < 1$.

3. $EG - F^2 > 0$ for $u^2 + v^2 < 1$, where

$$E = x_u^2 + y_u^2 + z_u^2,$$
$$F = x_u x_v + y_u y_v + z_u z_v,$$
$$G = x_v^2 + y_v^2 + z_v^2.$$

For many applications it would be important to replace 1. and 2. by some less restrictive condition. The least restrictive condition which can be substituted for 1. and 2., without hurting the validity of the conclusion, is not yet known[2]. From the point of view of the problem of PLATEAU, as it appears at the time being, condition 3. is however the most disturbing, inasmuch as we shall have to consider surfaces which possibly do not satisfy $EG - F^2 > 0$. It is easy to see that for such surfaces local isothermic parameters will *not* exist in general, no matter how good the coordinate functions are. Consider for instance the case

$$S: x = u, \quad y = 0, \quad z = 0, \quad u^2 + v^2 \leq 1.$$

Suppose we have local isothermic parameters. Using the same notations as at the beginning of the present section I.18, we have then

$$\iint\limits_{V_0} (EG - F^2)^{\frac{1}{2}} = \iint\limits_{V_0^*} (E^* G^* - F^{*2})^{\frac{1}{2}},$$

---

[1] L. LICHTENSTEIN: Beweis des Satzes etc. Abh. preuß. Akad. Wiss. 1911. — A. KORN: Zwei Anwendungen der Methode der sukzessiven Annäherungen (Schwarz-Festschrift), p. 215—229. Berlin, Julius Springer 1914.

* A function $f(u, v)$ satisfies in R the LIPSCHITZ-HÖLDER condition with the exponent λ if $|f(u_1, v_1) - f(u_2, v_2)| \leq \gamma d^\lambda$, where d is the distance of the points (u_1, v_1), (u_2, v_2) and γ is a constant.

[2] The reviewer had the privilege to get information about yet unpublished investigations of J. E. MCSHANE and of C. B. MORREY which lead to very general results. — In connection with his work on the problem of PLATEAU, T. RADÓ used maps which are conformal in a certain approximate sense and announced further developments on the subject [Bull. Amer. Math. Soc. Vol. 38 (1932) p. 129].

on account of the invariance of the area-integral. Since

$$E = 1, \quad F = 0, \quad G = 0, \quad E^* = G^*, \quad F^* = 0,$$

it follows that

$$\iint_{V_0^*} E^* = \iint_{V_0^*} G^* = 0,$$

and consequently $E^* \equiv 0$, $G^* \equiv 0$ in V_0^*. Hence $x_\alpha, y_\alpha, z_\alpha, x_\beta, y_\beta, z_\beta$ vanish identically in V_0^*, and thus x, y, z are constant in V_0^* and consequently in V_0, which however certainly is not the case.

I.19. As a substitute for the generally lacking conformal maps of surfaces which are not known to satisfy the condition $EG - F^2 > 0$, the fact that every polyhedron admits of a conformal map, in a certain generalized sense, has been used in the theory of the problem of PLATEAU[1]. We have namely the following theorem[2].

Every polyhedron $\mathfrak{P}$ admits of a typical representation (see I.9)

$$\mathfrak{P}: x = x(u, v), \quad y = y(u, v), \quad z = z(u, v), \quad u^2 + v^2 \leqq 1$$

with the following additional properties.

1. The sides of the (curvilinear) triangles $\delta_1, \delta_2, \ldots, \delta_n$ are analytic curves, including end-points.

2. None of the angles of these triangles is zero.

3. $x(u, v)$, $y(u, v)$, $z(u, v)$ are analytic in the interior of every one of the triangles $\delta_1, \delta_2, \ldots, \delta_n$ and satisfy there the equations $E = G$, $F = 0$.

Applications of this theorem will be considered in Chapter V and Chapter VI.

I.20. We shall review now certain facts, concerning curves, which will be used in the sequel. It is sufficient for our purposes to consider JORDAN arcs and JORDAN curves. A JORDAN arc is the one-to-one and continuous image of a closed interval $a \leqq t \leqq b$. A JORDAN curve is the one-to-one and continuous image of the unit circle $u^2 + v^2 = 1$.

Given two JORDAN arcs C_1 and C_2, their *distance* $d(C_1, C_2)$, in the sense of FRÉCHET, is defined as follows. Consider a topological correspondence T between C_1 and C_2, and denote by $M(T)$ the maximum distance of corresponding points. The greatest lower bound of $M(T)$, for all possible topological correspondences T, is by definition the distance $d(C_1, C_2)$. The definition of the distance of two JORDAN curves is exactly the same. Convergent sequences of JORDAN arcs and JORDAN curves are then defined in an obvious way in terms of the distance.

---

[1] See V.20.

[2] This theorem has been first proved by H. A. SCHWARZ: Über einen Grenzübergang durch alternierendes Verfahren. Gesammelte Mathematische Abhandlungen, Vol. II, pp. 133—143. The theorem is also comprised as a special case in the fundamental theorem of uniformisation. See for instance CARATHÉODORY, Conformal representation (Cambridge Tracts 28), Chapter VII.

It should be observed that a JORDAN arc and a JORDAN curve have no distance in the sense of FRÉCHET, since there is no topological correspondence between them to start with.

I.21. A JORDAN arc C, by definition, admits of a representation

$$C: x = x(t), \quad y = y(t), \quad z = z(t), \quad a \leqq t \leqq b,$$

where these equations define a one-to-one and continuous correspondence between $a \leqq t \leqq b$ and C. It will be necessary to consider improper representations also, such that while t describes the interval, the corresponding point moves on C always in the same sense, but possibly with stops and jumps. Since such a representation obviously does not determine C, we shall rather speak of monotonic transformations, all the more because the term points to the source of the significant properties we shall want. The following definitions will be used. Given a JORDAN arc C with end-points A^*, B^*, we say that the equations

$$x = x(t), \quad y = y(t), \quad z = z(t), \quad a \leqq t \leqq b$$

define a monotonic transformation of the interval $a \leqq t \leqq b$ into a set on C, if the following conditions are satisfied.

1. $t = a$ and $t = b$ are carried into A^* and B^* respectively.

2. If $a < t_1 < t_2 < b$, then the point P_1, corresponding to t_1, is on the arc bounded by A^* and the image P_2 of t_2, end-points included (if P_2 coincides with A^*, this implies that P_1 also coincides with A^*).

Instead of an interval and a JORDAN arc, we can consider two JORDAN arcs C_1, C_2 and speak of a monotonic transformation of C_1 into a set on C_2, the meaning being too obvious to be explained.

If C_1 and C_2 both reduce to straight segments, say to the intervals $a \leqq t \leqq b$ and $\alpha \leqq \tau \leqq \beta$, then the conditions 1., 2. clearly mean that $\tau(t)$ is a monotonic function such that $\tau(a) = \alpha$, $\tau(b) = \beta$ or $\tau(a) = \beta$, $\tau(b) = \alpha$. This situation suggests a number of simple theorems concerning monotonic transformations which generalize theorems on monotonic functions. An important theorem on monotonic functions is the theorem of HELLY, to the effect that a uniformly bounded sequence of monotonic functions always contains an everywhere convergent subsequence, the limit function being obviously monotonic again. Before we state the generalization, we extend the notion of monotonic transformations to JORDAN curves. Given a JORDAN curve Γ^*, and a set of equations

$$x = x(\Theta), \quad y = y(\Theta), \quad z = z(\Theta), \quad 0 \leqq \Theta < 2\pi,$$

we say that these equations define a monotonic transformation of the unit circle $u = \cos\Theta$, $v = \sin\Theta$ into a set on Γ^*, if the following conditions are satisfied.

1. There exist three distinct points A, B, C on the unit circle which are carried into three distinct points A^*, B^*, C^* of Γ^*.

2. The three non-overlapping arcs AB, BC, CA of the unit circle are transformed in a monotonic way into sets on the three non-overlapping arcs A^*B^*, B^*C^*, C^*A^* of Γ^*.

I.22. The theorem of HELLY admits then of the following obvious generalization. Let there be given a sequence Γ_n^* of JORDAN curves and for every one of them a monotonic transformation

$$T_n: x = x_n(\Theta), \quad y = y_n(\Theta), \quad z = z_n(\Theta)$$

of the unit circle $u = \cos\Theta$, $v = \sin\Theta$ into a set on Γ_n^*, such that three fixed distinct points A, B, C on the unit circle are carried into three distinct points A_n^*, B_n^*, C_n^* of Γ_n^*. Suppose the sequence Γ_n^* converges, in the FRÉCHET sense, to a JORDAN curve Γ^*, and also suppose that A_n^*, B_n^*, C_n^* converge to three distinct points A^*, B^*, C^* on Γ^*. Then it is possible to pick out a subsequence T_{n_k} such that the sequences $x_{n_k}(\Theta), y_{n_k}(\Theta), z_{n_k}(\Theta)$ converge everywhere on the unit circle. If the limit functions are denoted by $x(\Theta), y(\Theta), z(\Theta)$, then the equations $x = x(\Theta)$, $y = y(\Theta)$, $z = z(\Theta)$ define a monotonic transformation of the unit circle into a set on the limit curve Γ^*.

I.23. Let the equations

$$x = x(\Theta), \quad y = y(\Theta), \quad z = z(\Theta),$$

define a monotonic transformation T of the unit circle $u = \cos\Theta$, $v = \sin\Theta$ into a set on a JORDAN curve Γ^*. On account of the analogy with monotonic functions, there result the following facts for the functions $x(\Theta), y(\Theta), z(\Theta)$.

1. If Θ approaches from one side, that is to say in the counterclockwise or in the clockwise sense, any value Θ_0, then $x(\Theta), y(\Theta), z(\Theta)$ approach definite limits x_0^+, y_0^+, z_0^+ and x_0^-, y_0^-, z_0^- respectively.

2. If, for a certain Θ_0, we have $x_0^+ = x_0^-$, $y_0^+ = y_0^-$, $z_0^+ = z_0^-$, then $x(\Theta_0) = x_0^+ = x_0^-$, $y(\Theta_0) = y_0^+ = y_0^-$, $z(\Theta_0) = z_0^+ = z_0^-$, that is to say $x(\Theta), y(\Theta), z(\Theta)$ are then continuous for $\Theta = \Theta_0$.

3. If for two distinct points Θ_1, Θ_2 of the unit circle we have $x(\Theta_1) = x(\Theta_2)$, $y(\Theta_1) = y(\Theta_2)$, $z(\Theta_1) = z(\Theta_2)$, then $x(\Theta), y(\Theta), z(\Theta)$ all three reduce to constants on one of the two arcs determined by Θ_1 and Θ_2 on the unit circle.

I.24. While, in what precedes, we referred to the analogy with monotonic functions[1], J. DOUGLAS refers to results of FRÉCHET on parametric representations of the unit circle upon itself, to which the general case can be reduced. With every representation of the unit circle upon itself there can be associated a curve on a torus, and J. DOUGLAS obtains the necessary facts by a discussion of the situation on the torus[2].

[1] This point of view has been stressed by T. RADÓ: The problem of the least area and the problem of PLATEAU. Math. Z. Vol. 33 (1930) pp. 763—796

[2] See J. DOUGLAS: Solution of the problem of PLATEAU. Trans. Amer. Math. Soc. Vol. 33 (1931) pp. 270—272.

Chapter II.

Minimal surfaces in the small.

II.1. Given a surface

$$S: x = x(u, v), \quad y = y(u, v), \quad z = z(u, v),$$

it is convenient to use vector notation and write simply $\mathfrak{x} = \mathfrak{x}(u, v)$, where $\mathfrak{x}(u, v)$ denotes the vector with components $x(u, v), y(u, v), z(u, v)$. We have then for the first, second and third fundamental quantities of the surface the formulas[1]

$$E = \mathfrak{x}_u^2, \quad F = \mathfrak{x}_u \mathfrak{x}_v, \quad G = \mathfrak{x}_v^2,$$
$$L = \xi \mathfrak{x}_{uu}, \quad M = \xi \mathfrak{x}_{uv}, \quad N = \xi \mathfrak{x}_{vv},$$
$$e = \xi_u^2, \quad f = \xi_u \xi_v, \quad g = \xi_v^2.$$

ξ denotes the unit normal vector of the surface, the components X, Y, Z of which are given by the formulas

$$X = \frac{1}{W}\frac{\partial(y, z)}{\partial(u, v)}, \quad Y = \frac{1}{W}\frac{\partial(z, x)}{\partial(u, v)}, \quad Z = \frac{1}{W}\frac{\partial(x, y)}{\partial(u, v)},$$

where

$$W = (EG - F^2)^{\frac{1}{2}}.$$

For the total curvature K and for the mean curvature H respectively we have the formulas

$$K = \frac{LN - M^2}{W^2}, \quad H = \frac{EN - 2FM + GL}{2W^2}.$$

II.2. In writing all these formulas, we are making the obvious assumptions that $\mathfrak{x}(u, u)$ has the necessary differential coefficients and that $W > 0$. In differential geometry it is usually assumed that $\mathfrak{x}(u, v)$ is analytic. Many important issues are dodged by this latter blanket provision, particularly in the case of minimal surfaces in which we are interested. On the other hand, the assumption $W > 0$ is absolutely essential in differential geometry, since W appears in the denominator of practically every important formula.

It is convenient for our purposes to use the following definition. If $\mathfrak{x}(u, v)$ has continuous partial derivatives of the first order, and if $W > 0$, then the surface with the equation $\mathfrak{x} = \mathfrak{x}(u, v)$ will be called a regular surface of class C'. The classes $C'', C''', \ldots$ are defined in a similar way. The term *regular* refers to the condition $W > 0$. This condition secures the existence of the tangent plane. If we have occasion to change to new parameters, it is understood that we do this subject to the restriction that $\mathfrak{x}$ as function of the new parameters satisfies the same conditions.

[1] See for instance BLASCHKE: Vorlesungen über Differentialgeometrie. Vol. 1.

If $H \equiv 0$ for a regular surface S of class C'', then S is called a *minimal surface*. The term is misleading inasmuch as the area of a minimal surface, bounded by a given curve, is generally not a minimum (see III.14). A surface S, given by an equation $z = z(x, y)$, is a minimal surface if and only if $z(x, y)$ satisfies the partial differential equation

$$(1 + q^2)r - 2pqs + (1 + p^2)t = 0.$$

Indeed, the left-hand side is the numerator of the expression to which the general formula for H reduces for a surface given in the special representation $z = z(x, y)$.

II.3. It will be convenient to use the so-called differential notation in the sequel. If φ, ψ, ω are functions of u, v in a simply connected domain D, then the equation

$$d\omega = \varphi du + \psi dv$$

means that $\omega_u = \varphi$, $\omega_v = \psi$. We shall say then that $\varphi du + \psi dv$ is a complete differential in D, or that $\varphi du + \psi dv$ is the differential of ω. If φ, ψ have continuous partial derivatives of the first order, then $\varphi du + \psi dv$ is a complete differential if and only if $\varphi_v = \psi_u$.

If $\omega, \varphi_1, \varphi_2, \psi_1, \psi_2$ are functions of u, v in D, then the equation

$$d\omega = \varphi_1 d\varphi_2 + \psi_1 d\psi_2 \tag{2.1}$$

means that on substituting

$$d\varphi_2 = \varphi_{2u} du + \varphi_{2v} dv, \quad d\psi_2 = \psi_{2u} du + \psi_{2v} dv,$$

the right-hand side becomes the differential of ω in the sense explained above.

The convenience of the differential notation results from the fact that if new variables α, β are introduced by equations $\alpha = \alpha(u, v)$, $\beta = \beta(u, v)$, then $\omega, \varphi_1, \varphi_2, \psi_1, \psi_2$ considered as functions of α, β also satisfy the equation (2.1).

II.4. The study of minimal surfaces originated with the problem of the least area, that is to say the problem of determining and investigating a surface, bounded by a given curve, the area of which is a minimum[1].

Let Γ^* be the given boundary curve, and suppose there exists a surface S, bounded by Γ^*, the area of which is a minimum. Suppose also that S is a regular surface of class C''[2]. We ask then for consequences of the minimizing property of S, that is to say for necessary conditions for a minimum area. We are going to review briefly several of the ways in which this question has been handled in the older literature, restricting ourselves to cases which are significant from the point of view of the recent investigations.

[1] A beautiful presentation of the theory, on the basis of the older literature, is given in the classical work of Darboux: Théorie générale des surfaces. Vol. 1.

[2] The usual text-book assumptions are even more restrictive and therefore even less natural. Cf. II.12.

II.5. Choose a point P_0 on the minimizing surface S. If the axes x, y, z are properly chosen, then a sufficiently small vicinity S_0 of P_0 admits of a representation of the form $z = z(x, y)$, where $z(x, y)$ is single-valued in a certain region R_0 of the xy-plane and has continuous partial derivatives of the first and second order in R_0. From the minimizing property of S it follows that S_0 has also the minimizing property with respect to its own boundary curve. The area $\mathfrak{A}(S_0)$ of S_0 is given by

$$\mathfrak{A}(S_0) = \iint_{R_0} (1 + p^2 + q^2)^{\frac{1}{2}} \, dx \, dy .$$

Thus $z(x, y)$ minimizes this integral with respect to functions which coincide with $z(x, y)$ on the boundary of R_0.

Denote by $\lambda(x, y)$ a function which vanishes on the boundary of R_0 and has continuous derivatives of the first and second order in R_0. Let ε be a parameter. Then the area of the surface $z = z(x, y) + \varepsilon\lambda(x, y)$ is a function $A(\varepsilon)$ of ε which has a minimum for $\varepsilon = 0$. Hence

$$A'(0) = \iint_{R_0} \left(\frac{p}{W} \lambda_x + \frac{q}{W} \lambda_y\right) dx \, dy = 0 , \tag{2.2}$$

where

$$W = (1 + p^2 + q^2)^{\frac{1}{2}}.$$

Since (2.2) holds for all functions $\lambda(x, y)$ with the properties described above, it follows[1] that

$$\frac{\partial}{\partial x} \frac{p}{W} + \frac{\partial}{\partial y} \frac{q}{W} = 0 , \tag{2.3}$$

or

$$(1 + q^2) r - 2pqs + (1 + p^2) t = 0.$$

Hence (see II.2) the mean curvature of S_0 vanishes identically. As S_0 is the vicinity of an arbitrary point of S, it follows that the minimizing surface is a minimal surface.

II.6. The variation problem

$$\iint (1 + p^2 + q^2)^{\frac{1}{2}} \, dx \, dy = \text{minimum}$$

was the example which LAGRANGE considered to illustrate his method, yielding the so-called EULER-LAGRANGE equation, for the case of multiple integrals. As observed by LAGRANGE[2], the equation (2.3) means that

$$\frac{p \, dy - q \, dx}{(1 + p^2 + q^2)^{\frac{1}{2}}}$$

is a complete differential (cf. II.3).

II.7. Let again S be the minimizing surface, given in a general representation $\mathfrak{x} = \mathfrak{x}(u, v)$, where (u, v) varies in a region R. Let $\lambda(u, v)$ be a function which vanishes on the boundary of R and has

[1] See for instance BOLZA: Vorlesungen über Variationsrechnung, pp. 653—655.

[2] See DARBOUX: Théorie générale des surfaces. Vol. 1 pp. 267—268.

continuous derivatives of the first order in R, and let ε be a parameter. The area of the surface[1]

$$\mathfrak{x} = \mathfrak{x}(u, v) + \varepsilon\lambda(u, v)\,\xi(u, v),$$

where $\xi(u, v)$ is the unit normal vector of S, is then a function $A(\varepsilon)$ of ε which has a minimum for $\varepsilon = 0$. Hence $A'(0) = 0$, $A''(0) \geqq 0$. The first condition gives

$$\iint_R H W \lambda\, du\, dv = 0.$$

As this condition holds for every function λ with the properties specified above, it follows again that $H \equiv 0$, that is to say that S is a minimal surface. The condition $A''(0) \geqq 0$ gives then that

$$\iint_R \frac{1}{W}\{\lambda^2[Eg - 2Ff + Ge + 4(LN - M^2)] + E\lambda_v^2 - 2F\lambda_u\lambda_v + G\lambda_u^2\}\,du\,dv \geqq 0, \tag{2.4}$$

for all functions λ with the properties specified above. This condition permits us to construct examples of minimal surfaces which do not have the minimizing property (see III.14).

II.8. Let again the minimizing surface S be given in a general representation

$$S: x = x(u, v), \quad y = y(u, v), \quad z = z(u, v), \quad (u, v) \text{ in } R.$$

Let $\lambda(u, v)$ and ε have the same meaning as in II.7. Then the area of the surface

$$x = x(u, v) + \varepsilon\lambda(u, v), \quad y = y(u, v), \quad z = z(u, v)$$

is a function $A(\varepsilon)$ of ε which has a minimum for $\varepsilon = 0$. The condition $A'(0) = 0$ gives this time the equation[2]

$$\frac{\partial}{\partial u}\frac{Gx_u - Fx_v}{W} + \frac{\partial}{\partial v}\frac{Ex_v - Fx_u}{W} = 0. \tag{2.5}$$

Similar variations of the y and z coordinates respectively yield the further equations

$$\frac{\partial}{\partial u}\frac{Gy_u - Fy_v}{W} + \frac{\partial}{\partial v}\frac{Ey_v - Fy_u}{W} = 0, \tag{2.6}$$

$$\frac{\partial}{\partial u}\frac{Gz_u - Fz_v}{W} + \frac{\partial}{\partial v}\frac{Ez_v - Fz_u}{W} = 0. \tag{2.7}$$

In other words, the three expressions

$$\frac{Ex_v - Fx_u}{W}du - \frac{Gx_u - Fx_v}{W}dv, \tag{2.8}$$

$$\frac{Ey_v - Fy_u}{W}du - \frac{Gy_u - Fy_v}{W}dv, \tag{2.9}$$

$$\frac{Ez_v - Fz_u}{W}du - \frac{Gz_u - Fz_v}{W}dv \tag{2.10}$$

are complete differentials.

---

[1] See DARBOUX: Théorie générale des surfaces. Vol. 1 pp. 281–284.

[2] For the following formulas, see BOLZA: Vorlesungen über Variationsrechnung, p. 667.

II.9. The preceding results have important implications. Suppose that the minimizing surface S is given in terms of isothermic parameters[1]. Then $E = G$, $F = 0$, and the equations (2.5), (2.6), (2.7) reduce to

$$x_{uu} + x_{vv} = 0\,, \qquad y_{uu} + y_{vv} = 0\,, \qquad z_{uu} + z_{vv} = 0\,.$$

That is to say: *if the minimizing surface S is given in terms of isothermic parameters, then the coordinate functions $x(u, v)$, $y(u, v)$, $z(u, v)$ are harmonic functions.*

II.10. The preceding result may be obtained also by using the fact that a harmonic function $h(u, v)$ with given boundary values minimizes the DIRICHLET integral

$$\iint (h_u^2 + h_v^2)\, du\, dv\,.$$

Conversely, if a function $h(u, v)$ has this minimizing property, then $h(u, v)$ is harmonic. Consider then[2] the minimizing surface

$$S: \mathfrak{x} = \mathfrak{x}(u, v)\,, \qquad (u, v) \text{ in } R\,,$$

given in terms of isothermic parameters. On account of $E = G$, $F = 0$, we have then for the area $\mathfrak{A}(S)$ of S:

$$\mathfrak{A}(S) = \iint (EG - F^2)^{\frac{1}{2}} = \iint E = \iint G = \tfrac{1}{2} \iint (E + G)\,.$$

Denote by $\bar{\mathfrak{x}}(u, v)$ the vector whose components $\bar{x}(u, v)$, $\bar{y}(u, v)$, $\bar{z}(u, v)$ are the harmonic functions coinciding with the components $x(u, v)$, $y(u, v)$, $z(u, v)$ of $\mathfrak{x}(u, v)$ on the boundary of R, and denote by $\bar{S}$ the surface $\bar{S}: \mathfrak{x} = \bar{\mathfrak{x}}(u, v)$, (u, v) in R. On account of the minimizing property of harmonic functions we have then

$$\iint (\bar{x}_u^2 + \bar{x}_v^2) \leqq \iint (x_u^2 + x_v^2)\,, \tag{2.11}$$

$$\iint (\bar{y}_u^2 + \bar{y}_v^2) \leqq \iint (y_u^2 + y_v^2)\,, \tag{2.12}$$

$$\iint (\bar{z}_u^2 + \bar{z}_v^2) \leqq \iint (z_u^2 + z_v^2)\,, \tag{2.13}$$

and there follows by addition the inequality

$$\iint (\bar{E} + \bar{G}) \leqq \iint (E + G)\,.$$

On account of the minimizing property of S, we have $\mathfrak{A}(S) \leqq \mathfrak{A}(\bar{S})$. Summing up, we can write:

$$\mathfrak{A}(S) \leqq \mathfrak{A}(\bar{S}) = \iint (\bar{E}\bar{G} - \bar{F}^2)^{\frac{1}{2}} \leqq \iint \bar{E}^{\frac{1}{2}} \bar{G}^{\frac{1}{2}} \leqq \tfrac{1}{2} \iint (\bar{E} + \bar{G}) \leqq \tfrac{1}{2} \iint (E + G) = \mathfrak{A}(S)\,.$$

Consequently we must have the sign of equality all over, which implies the sign of equality in (2.11), (2.12), (2.13). That is to say, $x(u, v)$,

[1] See for instance BOLZA: Vorlesungen über Variationsrechnung, p. 667.

[2] The following reasoning is both a special case ($\sigma = 0$) and the origin of that used in V.20.

$y(u,v), z(u,v)$ minimize the DIRICHLET integral, and they are consequently harmonic functions.

II.11. Suppose the minimizing surface S is given by an equation $z = z(x, y)$. The expressions (2.8), (2.9), (2.10) reduce then to

$$-\frac{pq}{W}dx - \frac{1+q^2}{W}dy, \quad \frac{1+p^2}{W}dx + \frac{pq}{W}dy, \quad \frac{q}{W}dx - \frac{p}{W}dy, \tag{2.14}$$

where $W = (1 + p^2 + q^2)^{\frac{1}{2}}$. The components X, Y, Z of the unit normal vector are given by

$$X = -\frac{p}{W}, \quad Y = -\frac{q}{W}, \quad Z = \frac{1}{W}.$$

The expressions (2.14) are then found to be identical to

$$Y dz - Z dy, \quad Z dx - X dz, \quad X dy - Y dx.$$

Hence: for the minimizing surface S the expressions

$$Y dz - Z dy, \quad Z dx - X dz, \quad X dy - Y dx$$

are complete differentials[1].

II.12. The preceding properties of the minimizing surface S have been obtained under very restrictive assumptions concerning S. We shall see (III.13) that without proper restrictions concerning S the conclusion that S is a minimal surface is not in general valid.

There arises therefore the problem of determining the extent to which the restrictions imposed upon S can be lessened without hurting the validity of the conclusions obtained in the preceding sections. The discussion of this problem constitutes an essential part of most of the methods developed to handle the existence theorems in the problem of PLATEAU. The generality of the conclusions obtained in II.5, II.6, II.9 and II.11 will be considered in Chapter IV. The reasoning of II.10 will be generalized in Chapter V and Chapter VI in two different ways.

II.13. Since a minimal surface (as defined in II.2) generally does not have a minimum area, the question arises as to whether the conclusions obtained in the preceding sections for surfaces with a minimum area do or do not remain valid for minimal surfaces regardless of whether or not they have the minimizing property. This question leads in a very natural manner to the body of those theorems in the theory of minimal surfaces which are important for a clear understanding of the investigations reviewed in this report, and therefore we are going to discuss briefly that question.

II.14. *A regular surface S of class C'' is a minimal surface if and only if $Y dz - Z dy$, $Z dx - X dz$, $X dy - Y dx$ are complete differentials (X, Y, Z are the components of the unit normal vector of the surface).*

This theorem of H. A. SCHWARZ may be proved as follows. It clearly is sufficient to verify the theorem for small portions of S. Let

[1] Cf. II.14.

S_0 be a small simply connected portion of S. Then S_0 can be represented in one of the forms $z = z(x, y)$, $x = x(y, z)$, $y = y(z, x)$, say in the form $z = z(x, y)$. Then

$$X = -\frac{p}{W}, \quad Y = -\frac{q}{W}, \quad Z = \frac{1}{W}, \quad W = (1 + p^2 + q^2)^{\frac{1}{2}},$$

and

$$Y\,dz - Z\,dy = -\frac{pq}{W}\,dx - \frac{1+q^2}{W}\,dy, \tag{2.15}$$

$$Z\,dx - X\,dz = \frac{1+p^2}{W}\,dx + \frac{pq}{W}\,dy, \tag{2.16}$$

$$X\,dy - Y\,dx = \frac{q}{W}\,dx - \frac{p}{W}\,dy. \tag{2.17}$$

To apply the cross-wise differentiation test, we compute

$$\frac{\partial}{\partial y}\left(-\frac{pq}{W}\right) - \frac{\partial}{\partial x}\left(-\frac{1+q^2}{W}\right) = -\frac{p}{W^3}\,T,$$

$$\frac{\partial}{\partial y}\left(\frac{1+p^2}{W}\right) - \frac{\partial}{\partial x}\left(\frac{pq}{W}\right) = -\frac{q}{W^3}\,T,$$

$$\frac{\partial}{\partial y}\left(\frac{q}{W}\right) - \frac{\partial}{\partial x}\left(-\frac{p}{W}\right) = \frac{1}{W^3}\,T,$$

where

$$T = (1 + q^2)r - 2pqs + (1 + p^2)t.$$

These formulas show that the expressions (2.15), (2.16), (2.17) are complete differentials if and only if $T \equiv 0$, that is to say if and only if the surface is a minimal surface.

II.15. *If S is a minimal surface, and S_0 is a sufficiently small simply connected vicinity of an interior point P_0 of S, then S_0 can be mapped upon a plane region in a one-to-one and conformal way.*

It should be observed that since S is supposed to be of class C'', the theorem of I.18 is sufficiently general to apply. While reference to that theorem implies the use of the involved arguments necessary for its proof, the fact that S is a minimal surface enables us to prove in a direct and elementary way the existence of a conformal map[1].

The axes x, y, z being properly chosen, a sufficiently small simply connected vicinity S_0 of an interior point P_0 of S admits of a representation $z = z(x, y)$, where $z(x, y)$ is single-valued in a certain domain D_0 of the xy-plane and has there continuous partial derivatives of the first and second order. It is legitimate to suppose that D_0 is the interior of a circle. Since S_0 is a minimal surface, the expressions (2.15), (2.16), (2.17) in II.14 are complete differentials:

$$-\frac{pq}{W}\,dx - \frac{1+q^2}{W}\,dy = d\omega_1,$$

$$\frac{1+p^2}{W}\,dx + \frac{pq}{W}\,dy = d\omega_2,$$

$$\frac{q}{W}\,dx - \frac{p}{W}\,dy = d\omega_3,$$

[1] CH. H. MÜNTZ: Die Lösung des PLAUTEAUschen Problems über konvexen Bereichen. Math. Ann Vol. 94 (1925) pp. 53–96. — T. RADÓ: Über den analytischen Charakter der Minimalflächen. Math. Z. Vol. 24 (1925) pp. 321–327.

where $\omega_1, \omega_2, \omega_3$ are single-valued functions in D_0. Introduce then new variables α, β by the equations $\alpha = x$, $\beta = \omega_1(x, y)$. Since

$$\frac{\partial \beta}{\partial y} = -\frac{1+q^2}{W} < 0,$$

it is readily seen that D_0 is carried in a one-to-one and continuous way into a certain domain D_0^* of the $\alpha\beta$-plane. We obtain then the formulas

$$dx = d\alpha, \tag{2.18}$$

$$dy = -\frac{pq}{1+q^2}\,d\alpha - \frac{W}{1+q^2}\,d\beta, \tag{2.19}$$

$$dz = \frac{p}{1+q^2}\,d\alpha - \frac{qW}{1+q^2}\,d\beta, \tag{2.20}$$

$$d\omega_1 = d\beta, \tag{2.21}$$

$$d\omega_2 = \frac{W}{1+q^2}\,d\alpha - \frac{pq}{1+q^2}\,d\beta, \tag{2.22}$$

$$d\omega_3 = \frac{qW}{1+q^2}\,d\alpha + \frac{p}{1+q^2}\,d\beta. \tag{2.23}$$

The first three formulas yield $x_\alpha, x_\beta, y_\alpha, y_\beta, z_\alpha, z_\beta$, and direct computation shows that

$$x_\alpha^2 + y_\alpha^2 + z_\alpha^2 = \frac{1+p^2+q^2}{1+q^2} = x_\beta^2 + y_\beta^2 + z_\beta^2,$$

$$x_\alpha x_\beta + y_\alpha y_\beta + z_\alpha z_\beta = 0,$$

that is to say that α, β are isothermic parameters.

The lines $\alpha =$ constant of this conformal map correspond to the lines $x =$ constant on the surface. Hence the preceding computations prove the important theorem, discovered by RIEMANN and by BELTRAMI, that a minimal surface is intersected by parallel planes in curves which constitute an isothermic family.

II.16. The formulas (2.18) to (2.23) in II.15 show that

$$x_\alpha = \omega_{1\beta}, \quad x_\beta = -\omega_{1\alpha},$$

$$y_\alpha = \omega_{2\beta}, \quad y_\beta = -\omega_{2\alpha},$$

$$z_\alpha = \omega_{3\beta}, \quad z_\beta = -\omega_{3\alpha}.$$

That is to say, x and ω_1, y and ω_2, z and ω_3 satisfy, as functions of α and β, the CAUCHY-RIEMANN equations. It follows by inspection of the formulas defining these functions that they have continuous partial derivatives of the first order. Consequently, x, y, z as functions of α and β are harmonic and therefore analytic functions of α, β. Hence: *minimal surfaces are analytic*[1], although their definition (see II.2) implies only that they are of class C''. It follows, in particular, that

[1] CH. H. MÜNTZ: Die Lösung des PLATEAUschen Problems über konvexen Bereichen. Math. Ann. Vol. 94 (1925) pp. 53—96. — T. RADÓ: Über den analytischen Charakter der Minimalflächen. Math. Z. Vol. 24 (1925) pp. 321—327.

every solution $z(x, y)$, with continuous partial derivatives of the first and second order, of the equation

$$(1+q^2)r - 2pqs + (1+p^2)t = 0$$

is analytic.

II.17. The following theorem of WEIERSTRASS is fundamental for the theory of minimal surfaces.

Given a regular surface $S: \mathfrak{x} = \mathfrak{x}(u, v)$ of class C'' in terms of isothermic parameters. Then S is a minimal surface if and only if the components $x(u, v), y(u, v), z(u, v)$ of $\mathfrak{x}(u, v)$ are harmonic functions.

To prove this, show first that on account of $E = G$, $F = 0$ we have the identities:

$$Y dz - Z dy = x_v du - x_u dv,$$
$$Z dx - X dz = y_v du - y_u dv,$$
$$X dy - Y dx = z_v du - z_u dv.$$

Using the cross-wise differentiation test, we see that the expressions on the right-hand sides are complete differentials if and only if x, y, z are harmonic. On account of the theorem of SCHWARZ (see II.14), the expressions on the left-hand sides are complete differentials if and only if the surface is minimal. The theorem of WEIERSTRASS is an immediate consequence of these two facts.

II.18. The theorem of WEIERSTRASS leads to *standard formulas for minimal surfaces* which are fundamental for the existence theorems in the problem of PLATEAU.

According to II.15, a minimal surface admits in the small of isothermic parameters. That is to say, every minimal surface can be represented, in the small, by equations

$$x = x(u, v), \quad y = y(u, v), \quad z = z(u, v), \quad (u, v) \text{ in } D,$$

with $E = G, F = 0$. We can suppose that D is the interior of a circle. According to the theorem of WEIERSTRASS, $x(u, v), y(u, v), z(u, v)$ are then harmonic functions. We can write therefore:

$$x = \Re f_1(w), \quad y = \Re f_2(w), \quad z = \Re f_3(w),$$

where f_1, f_2, f_3 are analytic functions of the complex variable $w = u + iv$. It follows then that

$$x_u - i x_v = \varphi_1, \quad y_u - i y_v = \varphi_2, \quad z_u - i z_v = \varphi_3, \tag{2.24}$$

where $\varphi_1 = f_1'$, $\varphi_2 = f_2'$, $\varphi_3 = f_3'$ are again analytic functions of w. The condition $EG - F^2 > 0$, expressing that the surface is regular, is found to be equivalent to the condition that $\varphi_1, \varphi_2, \varphi_3$ do not vanish simultaneously at any point in D. Squaring and adding, we obtain from (2.24) the equation

$$E - G - 2iF = \varphi_1^2 + \varphi_2^2 + \varphi_3^2.$$

Thus $E = G, F = 0$ is equivalent to $\varphi_1^2 + \varphi_2^2 + \varphi_3^2 = 0$. This gives the theorem (discovered by MONGE):

Every minimal surface can be represented in the small by equations of the form

$$x = \Re \int^{w} \varphi_1 \, dw, \quad y = \Re \int^{w} \varphi_2 \, dw, \quad z = \Re \int^{w} \varphi_3 \, dw, \tag{2.25}$$

where $\varphi_1, \varphi_2, \varphi_3$ are analytic functions of w in the interior D of some circle and satisfying there the two conditions:

1. *$\varphi_1^2 + \varphi_2^2 + \varphi_3^2 = 0$ in D.*
2. *$\varphi_1, \varphi_2, \varphi_3$ do not vanish simultaneously at any point of D.*

Conversely, if $\varphi_1, \varphi_2, \varphi_3$ satisfy these conditions, then the equations (2.25) define a minimal surface.

II.19. The determination of the totality of minimal surfaces reduces therefore to the determination of all triples of functions $\varphi_1(w), \varphi_2(w), \varphi_3(w)$ with the properties 1. and 2. This can be done in several ways.

Choose a point w_0 in D. Since $\varphi_1(w_0), \varphi_2(w_0), \varphi_3(w_0)$ are not all three equal to zero, suppose that $\varphi_3(w_0) \neq 0$, for instance. We restrict ourselves to a vicinity D_0 of w_0, where $\varphi_3 \neq 0$. From $\varphi_1^2 + \varphi_2^2 + \varphi_3^2 = 0$ it follows that

$$(\varphi_1 - i\varphi_2)(\varphi_1 + i\varphi_2) = -\varphi_3^2. \tag{2.26}$$

Consequently $\varphi_1 - i\varphi_2 \neq 0$ in D_0. Put

$$\mu = \frac{\varphi_1 - i\varphi_2}{2}, \qquad \lambda = \frac{\varphi_3}{2\mu}. \tag{2.27}$$

Then λ, μ are analytic functions of w in D_0, and $\mu \neq 0, \lambda \neq 0$ in D_0. From the equations (2.26), (2.27) it follows then that

$$\varphi_1 = (1 - \lambda^2)\mu,$$
$$\varphi_2 = i(1 + \lambda^2)\mu,$$
$$\varphi_3 = 2\lambda\mu.$$

Conversely, if λ, μ are analytic functions of w, then the functions $\varphi_1, \varphi_2, \varphi_3$ satisfy the condition $\varphi_1^2 + \varphi_2^2 + \varphi_3^2 = 0$, as follows by direct computation. The formulas also show that if $\mu \neq 0$ in a domain D_0, then $\varphi_1, \varphi_2, \varphi_3$ do not vanish simultaneously at any point of D_0. Hence: the equations

$$\left.\begin{aligned} x &= \Re \int^{w} (1 - \lambda^2)\mu \, dw, \\ y &= \Re \int^{w} i(1 + \lambda^2)\mu \, dw, \\ z &= \Re \int^{w} 2\lambda\mu \, dw, \end{aligned}\right\} \tag{2.28}$$

where λ, μ are analytic functions of w in the interior D_0 of some circle and $\mu \neq 0$ in D_0, yield the totality of minimal surfaces, considered in the small.

II.20. In II.19, we supposed that $\varphi_3 \neq 0$ for instance. This situation always can be arranged for by renaming the axes x, y, z, if necessary. It should be observed that it might happen that it is absolutely necessary to rename the axes in order to represent a given minimal surface by the formulas (2.28). Consider for instance the equations

$$x = \Re \int^{w} \varphi_1 \, dw, \qquad y = \Re \int^{w} \varphi_2 \, dw, \qquad z = \Re \int^{w} \varphi_3 \, dw,$$

where $\varphi_1 = w^2 - 1$, $\varphi_2 = i(1 + w^2)$, $\varphi_3 = 2w$, in the vicinity of $w = 0$. Then $\varphi_1^2 + \varphi_2^2 + \varphi_3^2 = 0$, and $\varphi_1, \varphi_2, \varphi_3$ never vanish simultaneously. Still, $\varphi_1, \varphi_2, \varphi_3$ cannot be represented in the form (2.28) in the vicinity of $w = 0$. Indeed, from the equations

$$w^2 - 1 = (1 - \lambda^2)\mu, \qquad i(1 + w^2) = i(1 + \lambda^2)\mu, \qquad 2w = 2\lambda\mu$$

it would follow that $\mu = w^2$, $\lambda = \frac{1}{w}$. Thus λ would necessarily have a pole at $w = 0$, and μ would necessarily vanish at $w = 0$.

Thus the statement at the end of II.19 should be amended by saying that the formulas (2.28) yield all minimal surfaces provided we rename, if necessary, the axes x, y, z. This constitutes a very serious disadvantage of the formulas (2.28)[1].

II.21. A somewhat closer discussion shows that every triple $\varphi_1, \varphi_2, \varphi_3$ with the properties 1., 2. stated in II.18 can be represented, in the small, by the formulas

$$\varphi_1 = \Phi^2 - \Psi^2,$$
$$\varphi_2 = i(\Phi^2 + \Psi^2),$$
$$\varphi_3 = 2\Phi\Psi,$$

where Φ, Ψ are single-valued analytic functions of w which do not have any common zeros. This gives the *formulas of* WEIERSTRASS: every minimal surface can be represented, in the small, by equations

$$\left.\begin{aligned} x &= \Re \int^{w} (\Phi^2 - \Psi^2)\, dw, \\ y &= \Re \int^{w} i(\Phi^2 + \Psi^2)\, dw, \\ z &= \Re \int^{w} 2\Phi\Psi \, dw, \end{aligned}\right\} \tag{2.29}$$

where Φ, Ψ are single-valued analytic functions of w without common zeros[2]. The converse is obvious.

II.22. In later Chapters we shall have to consider minimal surfaces in a more general sense. The generalization will consist in dropping the

[1] A large part of the older work is based on the special case of (2.28) corresponding to $\lambda(w) = w$.

[2] The classical notation is G, H. As G, H have a standard meaning in differential geometry, we changed to Φ, Ψ.

condition of regularity $EG - F^2 > 0$ of the surface. On account of (2.24) this amounts to permitting the functions $\varphi_1, \varphi_2, \varphi_3$ of II.18 to vanish simultaneously. The question arises then if the formulas (2.29) still represent, in the small, all minimal surfaces. This is not the case, as might be inferred from the example

$$x = \Re \int^w \varphi_1 dw, \quad y = \Re \int^w \varphi_2 dw, \quad z = \Re \int^w \varphi_3 dw,$$

where $\varphi_1 = 3w$, $\varphi_2 = 5iw$, $\varphi_3 = 4w$, and the situation is considered in the vicinity of $w = 0$. Clearly $\varphi_1^2 + \varphi_2^2 + \varphi_3^2 = 0$. Suppose there would exist, in the vicinity of $w = 0$, two single-valued analytic functions Φ, Ψ such that $3w = \Phi^2 - \Psi^2$, $5iw = i(\Phi^2 + \Psi^2)$, $4w = 2\Phi\Psi$. It would follow that $\Phi^2 = 4w$, and it is then clear that Φ is not single-valued in the vicinity of $w = 0$.

An obvious remedy would be to permit the functions Φ, Ψ, in the formulas (2.29), to have algebraic singularities. As far as the investigations to be reviewed in the sequel are concerned, this step has not been taken (cf. Chapter V).

II.23. Since the purpose of this Chapter is to present only those facts about minimal surfaces which are significant from the point of view of the recent work on the problem of PLATEAU, we restrict ourselves to recall one more theorem, due to H. A. SCHWARZ[1]. *Suppose a minimal surface S contains a straight line g. Then g is an axis of symmetry of S.*

A simple proof may be obtained as follows. Choose the axes x, y, z in such a way that S may be represented, in the vicinity of one of its points situated on g, in the form $z = z(x, y)$, and that the y-axis coincides with g. Introduce then the isothermic parameters α, β defined in II.15. Since $z(0, y) \equiv 0$ and consequently $q(0, y) \equiv 0$ in the present case, it follows, with regard to the formulas in II.15, that

$$z = 0, \quad x = 0, \quad y_\alpha = 0 \text{ for } \alpha = 0. \tag{2.30}$$

Since x, y, z and consequently y_α, as functions of α and β, are harmonic functions (see II.16), it follows from (2.30) that

$$z(\alpha, \beta) = -z(-\alpha, \beta), \quad x(\alpha, \beta) = -x(-\alpha, \beta),$$
$$y_\alpha(\alpha, \beta) = -y_\alpha(-\alpha, \beta)$$

on account of the principle of symmetry. From the last of these three equations it follows by integration that $y(\alpha, \beta) = y(-\alpha, \beta)$. The equations $x(\alpha, \beta) = -x(-\alpha, \beta)$, $y(\alpha, \beta) = y(-\alpha, \beta)$, $z(\alpha, \beta) = -z(-\alpha, \beta)$ show that if (x_0, y_0, z_0) is a point of our minimal surface S, then $(-x_0, y_0, -z_0)$ is also a point of S. This proves the theorem of SCHWARZ.

[1] Gesammelte Mathematische Abhandlungen Vol. 1 p. 181.

Chapter III.

Minimal surfaces in the large.

III.1. The problem of PLATEAU, as considered for instance by H. A. SCHWARZ in his classical investigations[1], calls for a minimal surface $\mathfrak{M}$ which is bounded by a given Jordan curve Γ^* and which is *free of singularities in its interior*. The question naturally arises if this supplementary condition can be complied with if Γ^* is very complicated. The very particular cases which have been dealt with in the older literature certainly cannot be considered as representative of what might be expected in the case of a general JORDAN curve. At any rate, that supplementary condition has been dropped, at least for the time being, in the modern general investigations, and the existence of a solution free of singularities has been established only for certain special classes of curves.

III.2. If the solution is permitted to have singularities, it clearly is necessary to specify the nature of these singularities. For instance, if the solution would be permitted to have *edges*, then every polyhedron, bounded by a given polygon Γ^*, should be considered as a solution of the problem of PLATEAU for the contour Γ^*. Or consider a uniqueness theorem, for instance the theorem stated by H. A. SCHWARZ that the solution of the problem of PLATEAU is unique if the given contour is a skew quadrilateral[2]. H. A. SCHWARZ had only solutions in mind which are free of singularities. We shall see (III.12) that the theorem remains valid even if the solution is permitted to have certain singularities. On the other hand, the theorem breaks down if the specifications as to permissible singularities are too liberal, as follows from the remark made above.

III.3. The specifications as to the permissible singularities are by no means uniform in the recent literature. In other words, the different authors do not always consider the same problem. We first consider the most liberal specifications actually used in the literature.

Given a continuous surface $S: \mathfrak{x} = \mathfrak{x}(u, v)$, (u, v) in R, where R denotes a JORDAN region bounded by a JORDAN curve Γ, and given also a JORDAN curve Γ^* in the xyz-space, we shall say that S is a *minimal surface (of the type of the circular disc) bounded by* Γ^* if the following conditions are satisfied[3].

1. The equations of S carry Γ in a topological way into Γ^*.

---

[1] Gesammelte Mathematische Abhandlungen Vol. 1.

[2] Gesammelte Mathematische Abhandlungen Vol. 1 p. 111.

[3] T. RADÓ: Contributions to the theory of minimal surfaces. Acta Litt. Sci. Szeged Vol. 6 (1932) pp. 1—20.

2. For every point (u_0, v_0) interior to Γ there exists a vicinity V_0 and a one-to-one and continuous transformation $\alpha = \alpha_0(u, v)$, $\beta = \beta_0(u, v)$ of V_0 into some region $\overline{V}_0$ of the $\alpha\beta$-plane such that the components x, y, z of $\mathfrak{x}$ as functions of α, β are harmonic in $\overline{V}_0$ and satisfy there the relations

$$\overline{E} = \overline{G}, \quad \overline{F} = 0,$$

where $\overline{E} = \mathfrak{x}_\alpha^2$, $\overline{F} = \mathfrak{x}_\alpha \mathfrak{x}_\beta$, $\overline{G} = \mathfrak{x}_\beta^2$. Parameters α, β with this property will then be called local typical parameters for the vicinity of (u_0, v_0).

III.4. If $\overline{E}\overline{G} - \overline{F}^2 > 0$ at the point (α_0, β_0) into which (u_0, v_0) is carried, then the corresponding point of S will be called a *regular point*. The vicinity, on S, of a regular point is a minimal surface in the sense of differential geometry.

If $\overline{E}\overline{G} - \overline{F}^2 = 0$ at (α_0, β_0), then we have, on account of $\overline{E} = \overline{G}$, $\overline{F} = 0$, $x_\alpha = y_\alpha = z_\alpha = x_\beta = y_\beta = z_\beta = 0$ at (α_0, β_0). Suppose that all the partial derivatives of x, y, z with respect to α, β vanish at (α_0, β_0) up to and including a certain order $n \geqq 1$, while at least one of the derivatives of order $n + 1$ is different from zero. Then the point corresponding to (α_0, β_0) on S will be called a *branch-point* of order n.

The preceding notions are independent of the special choice of the local typical parameters α, β. If S has only regular points, then S as a whole is a minimal surface in the sense of differential geometry, and will be called a *regular minimal surface*.

III.5. We now shall state the problems which will be discussed in this report. First we have the following statements of the problem of PLATEAU, which have been considered actually in the literature.

Problem P_1. Given, in the xyz-space, a JORDAN curve Γ^*, determine a minimal surface, of the type of the circular disc, bounded by Γ^* (the term *minimal surface* being used in the sense of III.3).

Problem P_2. Solve problem P_1 under the supplementary condition that the solution admits of a representation $S: \mathfrak{x} = \mathfrak{x}(u, v)$, $u^2 + v^2 \leqq 1$, where the components $x(u, v)$, $y(u, v)$, $z(u, v)$ of $\mathfrak{x}(u, v)$ are continuous for $u^2 + v^2 \leqq 1$, harmonic for $u^2 + v^2 < 1$, and satisfy for $u^2 + v^2 < 1$ the equations $E = G$, $F = 0$. Furthermore, the equations $x = x(u, v)$, $y = y(u, v)$, $z = z(u, v)$ are required to carry $u^2 + v^2 = 1$ in a topological way into the given JORDAN curve Γ^*.

Problem P_3. Solve problem P_2 under the supplementary condition that the functions $x(u, v)$, $y(u, v)$, $z(u, v)$ admit of a representation of the form

$$x = \Re \int^w (\Phi^2 - \Psi^2)\, dw,$$

$$y = \Re \int^w i(\Phi^2 + \Psi^2)\, dw,$$

$$z = \Re \int^w 2\, \Phi \Psi\, dw,$$

where Φ and Ψ denote single-valued analytic functions of $w = u + iv$ in $|w| < 1$.

Problem P_4. Solve problem P_3 under the supplementary condition that Φ, Ψ do not have any common zero in $|w| < 1$.

Problem P_5. Suppose that the orthogonal projection of the given JORDAN curve Γ^* upon the xy-plane is a simply covered JORDAN curve Γ. Denote by R the JORDAN region bounded by Γ. Solve problem P_1 under the supplementary condition that the solution admits of a representation $S: z = z(x, y)$, (x, y) in R, where $z(x, y)$ is single-valued and continuous in R and analytic in the interior of R.

Besides the problem of PLATEAU, we shall have to consider *the problem of the least area*, which requires the determination of a continuous surface S bounded by a given JORDAN curve Γ^*, such that the LEBESGUE area $\mathfrak{A}(S)$ of S is a minimum if compared with the areas of all continuous surfaces bounded by Γ^*, it being understood that only surfaces of the type of the circular disc are considered[1].

Then comes *the simultaneous problem*, which requires the determination of a common solution of the problem of PLATEAU and of the problem of the least area[2]. According to the different statements of the problem of PLATEAU, we have, strictly speaking, a number of simultaneous problems. As a matter of fact, only the statements P_2 and P_5 of the problem of PLATEAU actually have been used in this connection.

Any one of the preceding problems gives rise to the question as to whether or not the solution is unique. Furthermore, there arises the question as to whether or not some of these problems are equivalent. The existence theorems concerned with the problems listed above will be discussed in subsequent Chapters. The other questions raised in this section will be considered in the present chapter. Sections III.6 to III.16 contain a number of special facts which are then coordinated in the sections III.17 to III.20.

III.6. The following definitions prove useful for the sequel. Let there be given, in a JORDAN region R of the uv-plane, a continuous function $g(u, v)$. Then $g(u, v)$ will be called a *generalized harmonic function* in R if for every interior point (u_0, v_0) of R the following condition is satisfied. There exists a vicinity V_0 of (u_0, v_0) and a topological transformation $u = u_0(\alpha, \beta)$, $v = v_0(\alpha, \beta)$ of V_0 into some region $\overline{V}_0$ of an $\alpha\beta$-plane, such that the function $g[u_0(\alpha, \beta), v_0(\alpha, \beta)]$ of α, β is harmonic in $\overline{V}_0$. Variables α, β with this property will be called *local typical variables* for the point (u_0, v_0).

[1] Surfaces of different topological types will be only considered at the end of Chapter VI.

[2] This problem has been called the problem of PLATEAU by LEBESGUE: Intégrale, longueur, aire. Ann. Mat. pura appl. Vol. 7 (1902) pp. 231—359.

Suppose $g(u, v)$ is a generalized harmonic function in R. Suppose that $g(u, v)$ vanishes at an interior point (u_0, v_0) of R. Let (α_0, β_0) be the image of (u_0, v_0) under the transformation $u = u_0(\alpha, \beta)$, $v = v_0(\alpha, \beta)$, where α, β are local typical variables for (u_0, v_0). Suppose that all the partial derivatives of g with respect to α, β vanish at (α_0, β_0) up to and including a certain order $(n - 1) \geqq 0$ while at least one of the partial derivatives of order n is different from zero. Then (u_0, v_0) will be called a zero of order n of $g(u, v)$. This notion is independent of the particular choice of the local typical variables α, β.

III.7. If $g_1(u, v)$, $g_2(u, v)$ are generalized harmonic functions in R, then $g_1 + g_2$ generally is not a generalized harmonic function. On the other hand, several important properties of harmonic functions remain valid for generalized harmonic functions. The property expressed by the principle of maximum and minimum clearly remains valid. The following lemma is an immediate consequence of well-known properties of harmonic functions which remain valid for generalized harmonic functions.

Lemma[1]. Let $g(u, v)$ be a generalized harmonic function in a (simply connected) JORDAN region R. Suppose that $g(u, v)$ has a zero (u_0, v_0) of order $n \geqq 1$ in the interior of R. Then $g(u, v)$ vanishes in at least $2n$ distinct points on the boundary of R.

III.8. Consider now a surface $S: \mathfrak{x} = \mathfrak{x}(u, v)$, (u, v) in R, which is a minimal surface (in the sense defined in III.3) bounded by a JORDAN curve Γ^*. If α, β are local typical parameters (see III.3) for an interior point (u_0, v_0) of R, then the components x, y, z of $\mathfrak{x}$ as functions of α, β are harmonic functions. Hence, if a, b, c, d are any four constants, then $ax + by + cz + d$ is also a harmonic function of α, β. Thus if a, b, c, d are any four constants, then the function $ax(u, v) + by(u, v) + cz(u, v) + d$ is a generalized harmonic function in R. The following theorems are immediate consequences of this remark, on account of the facts referred to in III.7.

1. If a convex region K in the xyz-space contains the boundary curve Γ^* of a minimal surface (see III.3) then the whole surface is contained in K†.

2. The tangent plane, at a regular point (see III.4) of the minimal surface, intersects the boundary curve in at least four distinct points[2].

---

[1] T. RADÓ: Contributions to the theory of minimal surfaces. Acta Litt. Sci. Szeged Vol. 6 (1932) p. 10, where the lemma is stated for $n = 2$.

† The reviewer learned about this theorem from L. FEJÉR.

[2] The theorem is also true for surfaces with negative curvature. This fact played an important role in the work of S. BERNSTEIN on partial differential equations of the elliptic type. See for references L. LICHTENSTEIN: Neuere Entwicklung usw. Enzyklopädie der math. Wiss. Vol. 2 (3) pp. 1277—1334.

3. Every plane passing through a branch-point of order n (see III.4) of the minimal surface intersects the boundary curve Γ^* in at least $2(n+1)$ distinct points[1].

III.9. The last theorem permits us to exclude the possibility of branch-points in certain cases. Suppose there exists, in the xyz-space, a straight line l such that no plane through l intersects the boundary curve Γ^* in more than two distinct points. Then the minimal surface cannot have branch-points, as follows immediately from theorem 3 in III.8. The assumption is, for instance, satisfied if Γ^* has a simply covered star-shaped JORDAN curve as its parallel or central projection upon some plane. If the projection has the stronger property of being convex, then a much stronger conclusion can be drawn. Suppose that the parallel projection of Γ^* upon some plane is a simply covered convex curve. Choose a plane perpendicular to the direction of projection for the xy-plane. Then the orthogonal projection of Γ^* upon the xy-plane is again a simply covered convex curve which we shall call Γ. Let

$$S: x = x(u, v), \quad y = y(u, v), \quad z = z(u, v), \quad (u, v) \text{ in } R$$

be the minimal surface under consideration. From theorem 3 in III.8 it follows that S has no branch-points. From theorem 2 in III.8 it follows that S has no tangent plane perpendicular to the xy-plane. From this it follows that S has a simply covered xy-projection in the small. Hence the equations $x = x(u, v)$, $y = y(u, v)$ define a transformation with the following properties.

1. The transformation is one-to-one and continuous in the vicinity of every interior point (u_0, v_0) of R.

2. The boundary of R is carried in a one-to-one and continuous way into the JORDAN curve Γ.

On account of the so-called monodromy theorem in topology[2], it follows then that the transformation

$$x = x(u, v), \; y = y(u, v), \; (u, v) \text{ in } R$$

carries R in a topological way into the JORDAN region bounded by Γ. Hence u, v can be expressed as single-valued continuous functions of x, y in R, and there follows then for the minimal surface S a representation

$$S: z = z(x, y), \quad (x, y) \text{ in or on } \Gamma,$$

where $z(x, y)$ is single-valued and continuous in and on Γ. Since S has no branch-points, S is a minimal surface in the sense of differential geometry. Hence (see II.16) $z(x, y)$ is analytic in the interior of Γ and satisfies there the partial differential equation

$$(1 + q^2) r - 2pqs + (1 + p^2) t = 0. \tag{3.1}$$

[1] See T. RADÓ: The problem of the least area and the problem of PLATEAU. Math. Z. Vol. 32 (1930) p. 794, where the theorem is stated for $n = 1$.

[2] See, for instance, KERÉKJÁRTÓ: Vorlesungen über Topologie I, p. 175.

Similar conclusions might be obtained if the boundary curve Γ^* is supposed to have a simply covered convex curve as its central projection upon some plane.

III.10. Summing up, we have the following results[1]. Let S be a minimal surface (in the sense of III.3) bounded by a JORDAN curve Γ^*. If Γ^* has a simply covered star-shaped JORDAN curve Γ as its parallel or central projection upon some plane, then S has no branch-points, that is to say S is a minimal surface in the sense of differential geometry. If the projection Γ is convex, then S does not intersect itself even in the large. If the orthogonal projection of Γ^* upon the xy-plane is a simply covered convex curve Γ, then S can be represented in the form $S: z = z(x, y)$, (x, y) in or on Γ, where $z(x, y)$ is single-valued and continuous in and on Γ, and satisfies in the interior of Γ the partial differential equation (3.1).

III.11. We are going to consider now certain *uniqueness theorems*. A first important fact in this connection is the uniqueness theorem for the partial differential equation (3.1). Let there be given, on a JORDAN curve Γ in the xy-plane, a continuous boundary function $\varphi(P)$ of the point P varying on Γ. If then $z_1(x, y)$, $z_2(x, y)$ are solutions of (3.1) which both reduce to $\varphi(P)$ on Γ, then $z_1(x, y) \equiv z_2(x, y)$ in the whole interior of Γ.[2]

Denote then by Γ^* the JORDAN curve, in the xyz-space, determined by the equation $z = \varphi(P)$. The above uniqueness theorem asserts that Γ^* cannot bound more than one minimal surface which has a simply covered xy-projection. This statement is rather unsatisfactory; indeed, we shall see (III.17) that the boundary curve of a minimal surface might very well have a simply covered xy-projection, while the minimal surface itself does not have this property. On the other hand, if the xy-projection of the boundary curve Γ^* is convex, then every minimal surface bounded by Γ^* has also a simply-covered xy-projection, on account of III.10. A similar argument holds in case Γ^* is known to have a simply-covered convex curve as its central projection. In this way results the following uniqueness theorem[3].

If a JORDAN *curve Γ^* has a simply covered convex curve as its parallel or central projection upon some plane, then Γ^* cannot bound more than*

[1] See T. RADÓ: The problem of the least area and the problem of PLATEAU. Math. Z. Vol. 32 (1930) pp. 763—796. — T. RADÓ: Contributions to the theory of minimal surfaces. Acta Litt. Sci. Szeged Vol. 6 (1932) pp. 1—20.

[2] See, also for references, the beautiful treatment of this theorem and of related subjects by E. HOPF: Elementare Bemerkungen über die Lösungen partieller Differentialgleichungen zweiter Ordnung vom elliptischen Typus. S.-B. preuß. Akad. Wiss. 1927 pp. 147—152. See also A. HAAR: Über reguläre Variationsprobleme. Acta Litt. Sci. Szeged Vol. 3 (1927) pp. 224—234.

[3] T. RADÓ: Contributions to the theory of minimal surfaces. Acta Litt. Sci. Szeged Vol. 6 (1932) pp. 1—20.

one minimal surface (this term being used in the general sense defined in III.3).

III.12. The assumptions of the preceding uniqueness theorem obviously are satisfied if Γ^* is a skew quadrilateral. For this case, the uniqueness theorem has been stated, without proof, by H. A. SCHWARZ. For the purpose of an application to be made later on we mention the following consequence of the uniqueness theorem. Suppose a JORDAN curve Γ^* is invariable under reflection upon a certain plane p. Every minimal surface, bounded by Γ^*, is carried by the reflection into a minimal surface bounded by Γ^*. Hence, if it is known that Γ^* bounds just one minimal surface S, then it follows that S is also invariable under the reflection. Repeated application of this remark to the case when Γ^* consists of the edges AB, BC, CD, DA of a regular tetrahedron with vertices A, B, C, D, leads to the result that the minimal surface bounded by Γ^* passes through the center of the tetrahedron (this fact has been verified by SCHWARZ by using the explicit formulas for the surface).

III.13. Let us consider now the relation between the problem of the least area and the simultaneous problem (see III.5). If a solution S of the problem of the least area satisfies the assumptions made in the classical Calculus of Variations, then S is a minimal surface (see II.5). Without those assumptions this conclusion in general does not hold. The following example is a slight modification of one given by LEBESGUE in his Thesis[1].

Let the given JORDAN curve Γ^* coincide with the circle

$$\Gamma^*: x^2 + y^2 = 1, \quad z = 0.$$

Then the area $\mathfrak{A}(S)$ of any continuous surface S (of the type of the circular disc) bounded by Γ^* is $\geqq \pi$. Consider then the surface

$$S: \begin{cases} x = 2^n (r - \frac{1}{2})^n \cos\varphi, & y = 2^n (r - \frac{1}{2})^n \sin\varphi, & z = 0 \text{ for } \frac{1}{2} \leqq r \leqq 1, \\ x = 0, & y = 0, & z = (1 - 4r^2)^n \text{ for } 0 \leqq r \leqq \frac{1}{2}, \end{cases}$$

where r, φ are polar coordinates in the uv-plane, and n is a given positive integer. Using the relations $u = r\cos\varphi$, $v = r\sin\varphi$, we have the equations of S appearing in the form

$$S: x = x(u, v), \quad y = y(u, v), \quad z = z(u, v), \quad u^2 + v^2 \leqq 1,$$

where $x(u, v)$, $y(u, v)$, $z(u, v)$ easily are seen to have continuous partial derivatives up to and including the order $n - 1$. S consists of the simply covered disc $x^2 + y^2 \leqq 1$, $z = 0$ and of the spine $x = 0$, $y = 0$, $0 \leqq z \leqq 1$. The area $\mathfrak{A}(S)$ is found to be equal to π, by computing the integral $\iint (EG - F^2)^{\frac{1}{2}}$. Hence the area of S is a minimum. Still, S is not a minimal surface, not even in the general sense defined in III.3. Indeed, a minimal surface is comprised in every convex region

[1] Intégrale, longueur, aire. Ann. Mat. pura appl. Vol. 7 (1902) pp. 231—359.

which contains its boundary curve (see III.8), and S obviously does not satisfy this condition.

Instead of using one spine as above, we can disfigure any given surface S by putting on it any finite number of spines, without changing its area. We shall see (Chapter VI) that the problem of the least area is solvable for every JORDAN curve. From these two facts it follows that *the problem of the least area has infinitely many solutions for every* JORDAN *curve* Γ^*, *and that a surface* S *which solves the problem is not in general a minimal surface.*

III.14. Thus the solution of the problem of the least area does not imply the solution of the problem of PLATEAU. Neither does the solution of the problem of PLATEAU imply the solution of the problem of the least area; in other words, *the area of a minimal surface, bounded by a given* JORDAN *curve* Γ^*, *is not necessarily a minimum.* This fact has been recognized at the earliest stage of the theory. For the case of doubly connected minimal surfaces bounded by two given curves, the catenoids offer simple examples for the lack of the minimizing property[1]. For the case of minimal surfaces, of the type of the circular disc, bounded by a given curve, H. A. SCHWARZ obtained very general examples in the following way[2].

Consider, in the $u + iv = w$ plane, a JORDAN region R bounded by an analytic JORDAN curve. Denote by $\mu(w)$ a function which is analytic and different from zero in R. Then the equations

$$\left.\begin{aligned} x &= \Re\int^{w}(1-w^2)\,\mu(w)\,dw\,,\\ y &= \Re\int^{w} i(1+w^2)\,\mu(w)\,dw\,,\\ z &= \Re\int^{w} 2w\,\mu(w)\,dw\,,\end{aligned}\right\} \tag{3.2}$$

where w varies in R, define a regular minimal surface[3]. The area of this surface certainly is not a minimum if the inequality (2.4) in II.7 is not satisfied. The left-hand side of that inequality reduces in the present case to

$$\iint\limits_{R}\left[\lambda_u^2+\lambda_v^2-\frac{8\lambda^2}{(1+u^2+v^2)^2}\right]du\,dv\,. \tag{3.3}$$

Hence the area of the minimal surface certainly is not a minimum if this integral can be made negative by substituting a function $\lambda(u, v)$ which has continuous partial derivatives of the first order in R and which vanishes on the boundary of R.

[1] See, also for references, the beautiful Chapter IV in the little book of G. A. BLISS: Calculus of Variations (No. 1 of the Carus Mathematical Monographs).

[2] Gesammelte Mathematische Abhandlungen Vol. 1 pp. 151—167 and 223—269.

[3] The formulas (3.2) are obtained by choosing $\lambda(w) = w$ in (2.28).

This criterion, curiously enough, does not depend upon the function $\mu(w)$ which determines the minimal surface; the criterion is concerned solely with the region R. SCHWARZ based the discussion of this situation on the study of the characteristic values of a certain partial differential equation[1]. One of the results he obtained states that if the region R contains the unit circle $u^2+v^2 \leqq 1$ in its interior, then the integral (3.3) can be made negative. On account of the geometrical meaning[2] of the variable w in the formulas (3.2), this assumption concerning R means that the spherical image of the minimal surface defined by (3.2) completely covers half the unit sphere.

This result of SCHWARZ can be obtained also in the following elementary way[3]. Let r be a positive parameter and define a function $\lambda(u, v; r)$ by

$$\lambda(u, v; r) = \frac{u^2+v^2-r^2}{u^2+v^2+r^2} \quad \text{for} \quad 0 \leqq u^2+v^2 \leqq r^2.$$

Put

$$J(r) = \iint\limits_{u^2+v^2<r^2} \left[\lambda_u^2 + \lambda_v^2 - \frac{8\lambda^2}{(1+u^2+v^2)^2}\right] du\, dv.$$

Partial integration gives

$$J(r) = -\iint\limits_{u^2+v^2<r^2} \lambda\left[\Delta\lambda + \frac{8\lambda}{(1+u^2+v^2)^2}\right] du\, dv.$$

Actual computation shows that the function $\lambda(u, v; 1)$ satisfies the equation

$$\Delta\lambda + \frac{8\lambda}{(1+u^2+v^2)^2} = 0.$$

Hence $J(1) = 0$. We want now to determine $J'(1)$. Changing to new variables

$$\alpha = \frac{u}{r}, \quad \beta = \frac{v}{r},$$

we transform $J(r)$ into a new integral taken over the fixed region $\alpha^2+\beta^2 < 1$. $J'(1)$ can then be computed by differentiating under the integral sign. The result is

$$J'(1) = 16 \iint\limits_{\alpha^2+\beta^2<1} \frac{(\alpha^2+\beta^2-1)^3}{(\alpha^2+\beta^2+1)^5}\, d\alpha\, d\beta.$$

Since the integrand obviously is negative, it follows that $J'(1) < 0$. From $J(1) = 0$, $J'(1) < 0$ it follows that we have a constant $\sigma > 1$, such that $J(r) < 0$ for $1 < r < \sigma$.

Suppose then that a JORDAN region R contains $u^2+v^2 \leqq 1$ in its interior. Then r can be determined in such a way that $1 < r < \sigma$ and

[1] Gesammelte Mathematische Abhandlungen Vol. 1 pp. 241–269.

[2] w is the stereographic projection of the spherical image of the surface. See DARBOUX: Théorie générale des surfaces Vol. 1 pp. 347–348.

[3] T. RADÓ: Contributions to the theory of minimal surfaces. Acta Litt. Sci. Szeged Vol. 6 (1932) pp. 1–20.

that the circle $u^2 + v^2 \leqq r^2$ is interior to R. Define a function $\lambda(u, v)$ by the formulas

$$\lambda(u, v) = \begin{cases} \lambda(u, v; r) & \text{in} \quad 0 \leqq u^2 + v^2 \leqq r^2, \\ 0 & \text{for} \quad u^2 + v^2 > r^2. \end{cases}$$

Then $\lambda(u, v)$ vanishes on the boundary of R and makes the integral (3.3) negative. The first partial derivatives of $\lambda(u, v)$ are discontinuous on $u^2 + v^2 = r^2$; this edge however can be rounded off by a familiar process.

It is thus proved that if R contains $u^2 + v^2 \leqq 1$ in its interior, then the formulas (3.2) define, for every choice of the analytic function $\mu(w)$, a minimal surface the area of which is not a minimum.

III.15. In order to obtain a clear-cut example, the function $\mu(w)$ in (3.2) should be chosen in such a way that the resulting minimal surface be bounded by a JORDAN curve. It can easily be shown that for $\mu(w) \equiv \frac{1}{3}$ this condition is satisfied, provided R is a circle $u^2 + v^2 \leqq r^2$ with $r < \sqrt{3}$. Then the following explicit example is obtained[1]. *The equations*

$$\left.\begin{aligned} x &= u + uv^2 - \tfrac{1}{3}u^3, \\ y &= -v - u^2 v + \tfrac{1}{3}v^3, \\ z &= u^2 - v^2, \end{aligned}\right\} \quad u^2 + v^2 \leqq r^2, \quad 1 < r < \sqrt{3}, \tag{3.4}$$

define a minimal surface S bounded by a JORDAN *curve, such that the area of S is not a minimum*[2].

III.16. The area of the minimal surface S given by (3.4) clearly is finite. On account of a general existence theorem to be considered in Chapter VI, it follows that the boundary curve of S bounds a minimal surface S^* the area of which is a minimum. Then S^* is not identical to S, because the area of S is not a minimum. Hence the boundary curve of S bounds at least two distinct minimal surfaces. The equations of the boundary curve, in terms of polar coordinates r, Θ, where $u = r\cos\Theta$, $v = r\sin\Theta$, are obtained in the form

$$\left.\begin{aligned} x &= r\cos\Theta - \tfrac{1}{3}r^3\cos 3\Theta, \\ y &= -r\sin\Theta - \tfrac{1}{3}r^3\sin 3\Theta, \\ z &= r^2\cos 2\Theta. \end{aligned}\right\} \quad 0 \leqq \Theta < 2\pi. \tag{3.5}$$

Thus we have the following example[1]: *if* $1 < r < \sqrt{3}$, *the equations (3.5) define a* JORDAN *curve which bounds at least two distinct minimal surfaces of the type of the circular disc.*

[1] T. RADÓ: Contributions to the theory of minimal surfaces. Acta Litt. Sci. Szeged Vol. 6 (1932) pp. 1—20.

[2] The formulas (3.4) define the so-called minimal surface of ENNEPER. See DARBOUX: Théorie générale des surfaces Vol. 1 pp. 372—376. Examples of a less elementary character have been given by H. A. SCHWARZ: Gesammelte Mathematische Abhandlungen Vol. 1 pp. 151—167 and 223—269.

The catenoids give explicit examples of distinct minimal surfaces with identical boundaries[1]. It seems that no explicit example has yet been given for two minimal surfaces of the type of the circular disc and bounded by the same JORDAN curve. In the above example, one of the two minimal surfaces is given explicitly by the formulas (3.4), while the other is only known to *exist*[2]. At any rate, *the solution of the problem of* PLATEAU *in general is not unique*; on the other hand, the solution is unique for certain special classes of curves (see III.11). It would be desirable to obtain more comprehensive information concerning those properties of the given boundary curve which determine the number of the solutions.

III.17. We shall now discuss briefly the statements of the problems P_1 to P_5, listed in III.5. While problem P_1 simply calls for a minimal surface S (of the type of the circle) bounded by a given JORDAN curve Γ^*, problems P_2 to P_5 require that S admits of a representation of a prescribed type. In the case of problem P_5, this implies a restriction on Γ^* also, namely the restriction that Γ^* admits of a simply covered xy-projection.

In the course of general investigations on partial differential equations of the elliptic type, S. BERNSTEIN[3] observed that a large class of problems, including as a special case our problem P_5, in general is not solvable. The following example[4] shows the geometrical reasons for this fact for problem P_5. Take a regular tetrahedron, with vertices A, B, C, D, and let the xy-plane pass through A, B, C. Denote by Γ^* the quadrilateral AB, BC, CD, DA. Then Γ^* has a simply covered xy-projection. Yet, problem P_5 does not have a solution for Γ^*. There certainly exists a minimal surface S bounded by Γ^*, as has been proved already by SCHWARZ. We also know (III.11) that S is the only minimal surface (of the type of the circle) bounded by Γ^*. S passes through the center O of the tetrahedron (III.12), and thus S contains two points, namely O and D, with the same xy-projection. Hence *there exists a unique minimal surface bounded by* Γ^*, *and this minimal surface does not satisfy the additional condition of having a simply covered xy-projection*[5].

---

[1] See G. A. BLISS: Calculus of Variations (No. 1 of the Carus Mathematical Monographs).

[2] The same remark applies to an example due to N. WIENER. See J. DOUGLAS: Solution of the problem of PLATEAU. Trans. Amer. Math. Soc. Vol. 33 (1931) p. 269.

[3] S. BERNSTEIN: Sur les équations du Calcul des Variations. Ann. École norm. Vol. 29 (1912) pp. 431—485. See in particular pp. 484—485.

[4] T. RADÓ: Contributions to the theory of minimal surfaces. Acta Litt. Sci. Szeged Vol. 6 (1932) pp. 1—20.

[5] The *uniqueness* is absolutely essential for the conclusion that problem P_5 is not solvable. For this reason, several examples presented in the literature are incomplete.

This example shows very clearly that the trouble comes from the fact that problem P_5 is a *non-parametric problem*, inasmuch as it calls for a surface represented by an equation of the form $z = z(x, y)$, instead of asking for a surface in the general parametric form

$$x = x(u, v), \quad y = y(u, v), \quad z = z(u, v).$$

III.18. The problems P_1 to P_4 (see III.5) are statements of *the problem of* PLATEAU *in the parametric form.* Problems P_2, P_3, P_4 call, roughly speaking, for a minimal surface in terms of *isothermic parameters in the large.* The character of this condition might be well illustrated in the special case when Γ^* is a plane curve.

Suppose Γ^* is in the xy-plane, and consider, for instance, a solution

$$S: x = x(u, v), \quad y = y(u, v), \quad z = z(u, v), \quad u^2 + v^2 \leqq 1$$

of problem P_2. Then $z(u, v) = 0$ on $u^2 + v^2 = 1$. As $z(u, v)$ is harmonic, it follows that $z(u, v) \equiv 0$. The conditions $E = G$, $F = 0$ reduce therefore to

$$x_u^2 + y_u^2 = x_v^2 + y_v^2, \quad x_u x_v + y_u y_v = 0.$$

This implies that $x(u, v)$ and $y(u, v)$ are conjugate harmonic functions, that is to say real and imaginary parts of an analytic function $f(w)$ of $w = u + iv$. The statement of problem P_2 requires that

1. $f(w)$ is analytic for $|w| < 1$, and
2. $f(w)$ is continuous for $|w| \leqq 1$, and
3. the equation $\zeta = f(w)$ carries $|w| = 1$ in a topological way into the JORDAN curve Γ^* of the $\zeta = x + iy$ plane.

On account of a classical theorem of DARBOUX (see V.19), this situation implies that the equation $\zeta = f(w)$ carries $|w| < 1$ in a one-to-one and conformal way into the interior of Γ^*.

Hence problem P_2 requires the mapping of the JORDAN region, bounded by Γ^*, in a one-to-one and continuous, and in the interior conformal, way upon the unit circle $|w| \leqq 1$. The same holds obviously for problems P_3 and P_4. While problems P_2, P_3, P_4 reduce, in the case of a curve situated in the xy-plane, to one of the fundamental problems in conformal mapping, problem P_1 simply is trivial in this case. Indeed, if R denotes the JORDAN region bounded by Γ^*, then the surface

$$S: x = x, \; y = y, \; z = 0, \; (x, y) \text{ in } R$$

clearly solves problem P_1.

It might be inferred from these remarks that problem P_1 is *easier* to solve than problem P_2. Curiously enough, nobody ever has tried to take advantage of this possibility. The problem of PLATEAU, in the parametric form, has always been asked in one of the statements P_2, P_3, P_4[1] which also require the mapping of the minimal surface conformally upon the unit circle in the large.

[1] Or in the even more severe form based on the formulas (3.2).

The question naturally arises whether or not every solution of problem P_1 is also a solution of problem P_2. This amounts to the question whether or not a minimal surface (in the general sense of III.3) bounded by a JORDAN curve Γ^* always admits of a representation as specified by the statement of problem P_2. It seems that the methods developed for dealing with the OSGOOD-CARATHÉODORY theorem[1] will permit, after proper adjustments, the answering of this question in the affirmative[2].

III.19. Consider now a solution

$$S: x = x(u, v), \quad y = y(u, v), \quad z = z(u, v), \quad u^2 + v^2 \leqq 1$$

of problem P_3 or P_4 (see III.5). We have then the equations

$$\left.\begin{aligned} x_u - i x_v &= \Phi^2 - \Psi^2, \\ y_u - i y_v &= i(\Phi^2 + \Psi^2), \\ z_u - i z_v &= 2\Phi\Psi. \end{aligned}\right\} \tag{3.6}$$

The condition that $EG - F^2 = 0$ for an interior point (u_0, v_0) is then readily found to be equivalent to $\Phi = 0$, $\Psi = 0$ for $w = w_0 = u_0 + i v_0$. Since problem P_4 requires that Φ and Ψ have no common zeros, we see that *in problem P_4 the solution is not permitted to have branch-points.* It easily is seen that if a solution of problem P_2 does not have branch-points, then it admits of a representation as required in problem P_4. In other words, if we add the condition $EG - F^2 > 0$ for $u^2 + v^2 < 1$ in the statement of problem P_2, then we obtain problem P_4, that is to say the classical statement of the problem of PLATEAU in the parametric form[3].

The question arises whether or not problem P_4 is always possible. The impossibility, in general, of the problem would be demonstrated by exhibiting a single JORDAN curve Γ^* for which it could be proved that *every* minimal surface (of the type of the circular disc) bounded by Γ^* necessarily has branch-points. Such a curve has not yet been exhibited; the surmisal that any knotted JORDAN curve would serve the purpose can readily be refuted by examples of knotted JORDAN curves which do bound minimal surfaces (of the type of the circular disc) free of branch-points. More explicitly: there exist knotted JORDAN curves for which the classical problem P_4 is possible, and no JORDAN curve is known at present for which problem P_4 is impossible[4].

---

[1] We mean the following theorem: if the interior of a JORDAN curve is mapped, in a one-to-one and conformal way, upon the interior of the unit circle, then the map remains continuous and one-to-one on the boundaries. See for instance CARATHÉODORY: Conformal representation (Cambridge Tracts 28).

[2] This program has been carried out in a joint paper by E. F. BECKENBACH and T. RADÓ: Subharmonic functions and minimal surfaces. To appear in Trans. Amer. Math. Soc.

[3] See H. A. SCHWARZ: Gesammelte Mathematische Abhandlungen Vol. I.

[4] Cf. VI.35.

Combining the existence theorem in Chapter V with the theorems in III.10, we obtain existence theorems for the classical problem P_4 which seem to be the most general known at present. For instance: if a JORDAN curve Γ^* has a simply covered star-shaped curve as its parallel or central projection upon some plane, then problem P_4 is solvable for Γ^*.

III.20. While problem P_4 excludes branch-points altogether, and while problem P_2 does not imply any restriction as to branch-points, the geometrical interpretation of problem P_3 is less clear-cut.

Differentiating again the equations (3.6), we obtain

$$\left.\begin{aligned} x_{uu} - i x_{uv} &= 2(\Phi\Phi' - \Psi\Psi'), \\ y_{uu} - i y_{uv} &= 2i(\Phi\Phi' + \Psi\Psi'), \\ z_{uu} - i z_{uv} &= 2(\Phi\Psi' + \Phi'\Psi). \end{aligned}\right\} \tag{3.7}$$

From (3.6) and (3.7) it follows readily: if, for a solution of problem P_3, we have $EG - F^2 = 0$ at an interior point $u_0 + iv_0 = w_0$, then the partial derivatives of the first and second order of $x(u, v)$, $y(u, v)$, $z(u, v)$ all vanish at that point. In other words: *the branch-points of a solution of problem P_3 are at least of order* 2 (see III.4 for the definition). On the other hand, examples show that problem P_2 might have solutions with branch-points of order one.

Thus it follows that *the problems P_2 and P_3 are not equivalent.* Problem P_3 will be seen to be solvable for every not knotted JORDAN curve (Chapter V). Again, it is not known at the present time whether there do or do not exist curves for which problem P_3 is impossible.

III.21. We have seen (III.18) that if the given JORDAN curve Γ^* is situated in the xy-plane, then the problem of PLATEAU in the parametric form (in any one of the statements P_2, P_3, P_4 of III.5) reduces to the problem of mapping the JORDAN region bounded by Γ^* in a one-to-one and continuous, and in the interior conformal, way upon the unit circle $u^2 + v^2 \leqq 1$. This situation played an important role in the theory. The most direct illustration is given by the case when Γ^* is a *polygon.* One of the earliest ideas for dealing with the problem of the conformal mapping of the unit circle upon the region bounded by a plane polygon was based on the principle of symmetry and led to the so-called formulas of SCHWARZ and CHRISTOFFEL. The method used by SCHWARZ in his classical investigations on the problem of PLATEAU is clearly a generalization to the space of the plane method[1]. In a general way, the reader will find, in the Chapters IV, V, VI dealing with the existence theorems, many instances where the theory of the conformal mapping of plane

[1] Compare, for instance, the following two papers of H. A. SCHWARZ: Über einige Abbildungsaufgaben. Gesammelte Mathematische Abhandlungen Vol. 2 pp. 65—83 and Bestimmung einer speziellen Minimalfläche. Gesammelte Mathematische Abhandlungen Vol. 1 pp. 6—125. See also DARBOUX: Théorie générale des surfaces Vol. 1 pp. 490—601.

regions clearly served as a model for the development of the theory of the problem of PLATEAU.

An important remark, due to J. DOUGLAS, should be mentioned here. Since the problem of OSGOOD-CARATHÉODORY (conformal mapping of a plane JORDAN region upon the circle) is included in problem P_2, it follows that if we have a method of solution of problem P_2 which does not make use of the solution of that problem, then we have a simultaneous solution of the problem of OSGOOD-CARATHÉODORY and of the problem of PLATEAU. J. DOUGLAS emphasizes the fact that this is the case with his own method[1].

III.22. Instead of considering only existence theorems, the relation between conformal mapping of plane regions, that is to say analytic functions of a complex variable, on the one side and minimal surfaces on the other side might be discussed on account of its own intrinsic interest. The theory of minimal surfaces appears then as a generalization of the theory of analytic functions of a complex variable. While the Chapters IV, V, VI will review numerous facts and methods which might be interpreted from this point of view, it might be useful to present to the reader some specific illustrations.

III.23. If $f(w) = x(u, v) + iy(u, v)$ is an analytic function of w for $|w|<1$, then $x_u = y_v$, $x_v = -y_u$ (CAUCHY-RIEMANN equations). From these equations it follows that

$$x_u^2 + y_u^2 = x_v^2 + y_v^2, \qquad x_u x_v + y_u y_v = 0. \tag{3.8}$$

Conversely, it follows from (3.8) that either y is conjugate harmonic to x, or x is conjugate harmonic to y. Let us call two harmonic functions related by (3.8) *a couple of conjugate harmonic functions*.

On the other hand, the theorem of WEIERSTRASS (see II.17) leads to consider triples of harmonic functions related by the equations

$$x_u^2 + y_u^2 + z_u^2 = x_v^2 + y_v^2 + z_v^2, \qquad x_u x_v + y_u y_v + z_u z_v = 0. \tag{3.9}$$

We shall say that x, y, z form *a triple of conjugate harmonic functions*. We shall review now a few facts which develop further this analogy.

III.24. Suppose $f(w)$ is analytic in $|w|<1$ and even on $|w|=1$, for the sake of simplicity. Put $f'(w) = g(w)$. Then the area of the image of $|w|\leq 1$ by $f(w)$ is given by

$$\mathfrak{A} = \int_0^1\int_0^{2\pi} |g(re^{i\Theta})|^2 r\,dr\,d\Theta, \tag{3.10}$$

while the length L of the image of $|w|=1$ is

$$L = \int_0^{2\pi} |g(e^{i\Theta})|\,d\Theta. \tag{3.11}$$

[1] See J. DOUGLAS: Solution of the problem of PLATEAU. Trans. Amer. Math. Soc. Vol. 33 (1931) pp. 263—321.

If $f(w)$ would give a simply covered image of $|w| \leqq 1$, then we could assert, on account of the isoperimetric inequality, that

$$\mathfrak{A} \leqq \frac{1}{4\pi} L^2,$$

or, on account of (3.10) and (3.11), that

$$\int_0^1 \int_0^{2\pi} |g(re^{i\Theta})|^2 r\,dr\,d\Theta \leqq \frac{1}{4\pi} \left(\int_0^{2\pi} |g(e^{i\Theta})|\,d\Theta \right)^2. \tag{3.12}$$

Carleman[1] proved that (3.12) holds regardless if $f(w)$ gives a simply covered image. From this he inferred that between the area $\mathfrak{A}$ of a minimal surface and the length L of its boundary curve the isoperimetric inequality also holds. To prove this[2], suppose, to simplify the discussion, that the minimal surface is given by

$$S: x = x(u, v), \quad y = y(u, v), \quad z = z(u, v), \quad u^2 + v^2 \leqq 1, \tag{3.13}$$

where x, y, z form a triple of conjugate harmonic functions (see III.23) which remain analytic even on $u^2 + v^2 = 1$. Then x, y, z are the real parts of analytic functions:

$$x = \Re f_1(w), \quad y = \Re f_2(w), \quad z = \Re f_3(w),$$

and if we put $f_1' = g_1$, $f_2' = g_2$, $f_3' = g_3$ and use the equation $g_1^2 + g_2^2 + g_3^2 = 0$ (see II.18), then the isoperimetric inequality is expressed by

$$\frac{1}{2} \sum_{k=1}^{3} \int_0^1 \int_0^{2\pi} |g_k(re^{i\Theta})|^2 r\,dr\,d\Theta \leqq \frac{1}{4\pi} \left[\int_0^{2\pi} \left(\frac{1}{2} \sum_{k=1}^{3} |g_k(e^{i\Theta})|^2 \right)^{\frac{1}{2}} d\Theta \right]^2. \tag{3.14}$$

To prove (3.14), observe that on account of the inequality of Minkowski[3] we have

$$\frac{1}{2} \sum_{k=1}^{3} \left(\int_0^{2\pi} |g_k(e^{i\Theta})|\,d\Theta \right)^2 \leqq \left[\int_0^{2\pi} \left(\frac{1}{2} \sum_{k=1}^{3} |g_k(e^{i\Theta})|^2 \right)^{\frac{1}{2}} d\Theta \right]^2. \tag{3.15}$$

Hence (3.14) follows immediately from (3.15) and (3.12). Some more discussion shows that the sign of equality in (3.14) holds if and only if the minimal surface reduces to a simply covered circular disc.

If we suppose that the minimal surface has a minimum area, then the theorem of Carleman is almost trivial, as it has been observed by Blaschke[4]. Indeed, consider a cone consisting of the straight segments which connect a fixed point of the boundary curve with a variable point on that curve. Since cones are developable, the isoperimetric inequality holds for this cone, and hence all the more for the minimal surface, since the area of this latter is by assumption not greater than the area

[1] Zur Theorie der Minimalflächen. Math. Z. Vol. 9 (1921) pp. 154—160.

[2] We follow the simple proof given by E. F. Beckenbach: The area and boundary of minimal surfaces. Ann. of Math. Vol. 33 (1932) pp. 658—664.

[3] See for instance Pólya-Szegö: Aufgaben und Lehrsätze Vol. 2 p. 14.

[4] T. Carleman: Zur Theorie der Minimalflächen. Math. Z. Vol. 9 (1921) p. 160.

of the cone, while the boundary curve is the same for either surface. This remark of BLASCHKE shows that the point to the theorem of CARLEMAN is that the theorem is true even if the area of the minimal surface is not a minimum (cf. III.14).

III.25. Further inequalities between the area $\mathfrak{A}$ of a minimal surface and the length L of its boundary curve have been obtained by BECKENBACH[1]. Suppose that the minimal surface is again given by the equations (3.13). Suppose also that at the origin we have $E = 1$ (that is to say, the linear ratio of magnification at the origin is unity). Then

$$\mathfrak{A} \geqq \pi, \qquad L \geqq 2\pi.$$

That is to say, $\mathfrak{A}$ is at least equal to the area and L is at least equal to the perimeter of the unit circle. The sign of equality holds if and only if the minimal surface is a simply covered circular disc. These and similar theorems are proved by BECKENBACH by using FOURIER expansions.

III.26. A minimal surface S being again given by (3.13) where x, y, z are supposed to form a triple of conjugate harmonic functions, we shall call $(x^2 + y^2 + z^2)^{\frac{1}{2}}$ the *norm* of S and we shall write $(x^2 + y^2 + z^2)^{\frac{1}{2}} = |S|$. Then $|S|$ is the generalization of the absolute value of an analytic function $f(w)$ of w. If $f(w)$ is analytic, then $\log|f(w)|$ is a harmonic function of u, v, and this accounts for many important facts in the theory of functions of a complex variable. While $\log|S|$ is not harmonic, it can be shown to be *subharmonic*[2]. A function $g(u, v)$ is subharmonic in a domain D if for every point (u_0, v_0) of D the inequality

$$g(u_0, v_0) \leqq \frac{1}{2\pi} \int_0^{2\pi} g(u_0 + \varrho \cos\Theta, v_0 + \varrho \sin\Theta)\, d\Theta$$

is satisfied for sufficiently small values of ϱ[3]. If g has continuous partial derivatives of the second order, then this condition is equivalent to $\Delta g = g_{uu} + g_{vv} \geqq 0$ and $\log|S|$ is easily shown to satisfy this latter condition. It is sufficient to consider the situation in domains D where $|S| > 0$. Put $\log|S| = g$. Direct computation gives

$$\Delta g = \frac{(\mathfrak{x}_u^2 + \mathfrak{x}_v^2)\,\mathfrak{x}^2 - 2\left((\mathfrak{x}\mathfrak{x}_u)^2 + (\mathfrak{x}\mathfrak{x}_v)^2\right)}{|S|^4}, \tag{3.16}$$

[1] See E. F. BECKENBACH: The area and boundary of minimal surfaces. Ann. of Math. Vol. 33 (1932) pp. 658—664. — The theorems in III.24 and III.25 also hold for surfaces of negative curvature, given in isothermic representation. See E. F. BECKENBACH and T. RADÓ: Subharmonic functions and surfaces of negative curvature. To appear in Trans. Amer. Math. Soc.

[2] III.26 to III.29 are taken from E. F. BECKENBACH and T. RADÓ: Subharmonic functions and minimal surfaces. To appear in Trans. Amer. Math. Soc.

[3] See F. RIESZ: Sur les fonctions subharmoniques etc. Acta Math. Vol. 48 (1926) pp. 329—343.

where $\mathfrak{x} = \mathfrak{x}(u, v)$ is the vector equation of S. Since the components of $\mathfrak{x}$ form a triple of conjugate harmonic functions, we have

$$\mathfrak{x}_u^2 = \mathfrak{x}_v^2, \quad \mathfrak{x}_u \mathfrak{x}_v = 0.$$

At those points where $\mathfrak{x}_u^2 = \mathfrak{x}_v^2 = 0$, we have $\Delta g = 0$. At those points where $\mathfrak{x}_u^2 = \mathfrak{x}_v^2 > 0$, $\mathfrak{x}_u$ and $\mathfrak{x}_v$ are different from zero and are perpendicular to each other. Using a unit vector ξ perpendicular to $\mathfrak{x}_u$ and $\mathfrak{x}_v$, we can write

$$\mathfrak{x} = a\mathfrak{x}_u + b\mathfrak{x}_v + c\xi,$$

where a, b, c are scalars. It follows that

$$\mathfrak{x}\mathfrak{x}_u = a\lambda, \quad \mathfrak{x}\mathfrak{x}_v = b\lambda, \quad \mathfrak{x}^2 = \lambda(a^2 + b^2) + c^2, \quad \text{where} \quad \lambda = \mathfrak{x}_u^2 = \mathfrak{x}_v^2.$$

Substituting in (3.16), we get

$$\Delta g = \frac{2\lambda c^2}{|S|^4} \geqq 0.$$

Thus $\Delta g \geqq 0$ always.

III.27. The fact that $\log|S|$ is subharmonic makes it possible to extend a great number of theorems on analytic functions to minimal surfaces, namely those theorems which depend essentially on the fact that the product of analytic functions is again an analytic function. While there is no direct analogy for minimal surfaces, the subharmonic character of $\log|S|$ permits to extend the proofs. We just mention two examples.

Let the minimal surface S be given by the equations (3.13) where x, y, z form a triple of conjugate harmonic functions. Suppose that

1. $x(0,0) = 0, \quad y(0,0) = 0, \quad z(0,0) = 0,$
2. $(x^2 + y^2 + z^2)^{\frac{1}{2}} = |S| \leqq 1 \quad \text{in} \quad u^2 + v^2 < 1.$

Then

$$(x(u,v)^2 + y(u,v)^2 + z(u,v)^2)^{\frac{1}{2}} \leqq (u^2 + v^2)^{\frac{1}{2}} \quad \text{in} \quad u^2 + v^2 < 1,$$

and the sign of equality holds if and only if the surface is a simply covered circular disc. This generalizes the lemma of SCHWARZ.

III.28. Suppose this time that the minimal surface is given by

$$S: x = x(u,v), \quad y = y(u,v), \quad z = z(u,v), \quad 0 < \operatorname{arctg}\frac{v}{u} < \alpha,$$

where x, y, z form a triple of conjugate harmonic functions. Suppose these functions remain continuous for $v = 0$, $u > 0$, and suppose that $x(u,0)$, $y(u,0)$, $z(u,0)$ approach definite finite limits x_0^+, y_0^+, z_0^+ for $u \to +0$. Then, in every angle $0 \leqq \operatorname{arctg}\frac{v}{u} \leqq \alpha - \varepsilon$, $\varepsilon > 0$, the functions $x(u,v)$, $y(u,v)$, $z(u,v)$ approach the limits x_0^+, y_0^+, z_0^+ for $(u,v) \to (0,0)$. This generalizes a well-known theorem of LINDELÖF. The proof follows by an extension of the so-called *multiplication method*[1].

[1] See for instance PÓLYA-SZEGÖ: Aufgaben und Lehrsätze Vol. 1 p. 138 problem 277.

III.29. The theorem of III.26 can be completed as follows. *Three functions* $x(u, v)$, $y(u, v)$, $z(u, v)$, *continuous in a domain* D, *form there a triple of conjugate harmonic functions if and only if* $\log[(x+a)^2 + (y+b)^2 + (z+c)^2]$ *is subharmonic for every choice of the constants* a, b, c.

Chapter IV.

The non-parametric problem.

IV.1. The problem of PLATEAU in the non-parametric form asks for a minimal surface bounded by a given curve and represented as a whole by an equation of the form $z = z(x, y)$. The exact statement of the problem is as follows. Given, in the xy-plane, a JORDAN curve Γ, and a continuous function $\varphi(P)$ of the point P varying on Γ. Determine a function $z(x, y)$ which is continuous in and on Γ, which has continuous derivatives of the first and second orders inside of Γ, which reduces to $\varphi(P)$ on Γ, and which satisfies inside of Γ the partial differential equation

$$(1 + q^2) r - 2pqs + (1 + p^2) t = 0,$$

where

$$p = z_x, \quad q = z_y, \quad r = z_{xx}, \quad s = z_{xy}, \quad t = z_{yy}.$$

IV.2. It has been observed (III.17) that this problem, in general, has no solution. The methods developed for the solution of the problem therefore necessarily are based on some further assumptions concerning the prescribed boundary conditions. In the work we are going to review in this chapter, it is supposed that

1. the curve Γ which bears the given boundary values is convex, and
2. the boundary function $\varphi(P)$ satisfies a *three-point condition.*

The three-point condition may be stated as follows. Let Γ^* be the JORDAN curve defined in the xyz-space by the equation $z = \varphi(P)$. Let P_1^*, P_2^*, P_3^* be three distinct points on Γ^*, and denote by Θ the positive acute angle between the xy-plane and the plane passing through P_1^*, P_2^*, P_3^*. If, for all possible positions of the points P_1^*, P_2^*, P_3^*, the quantity $\operatorname{tg}\Theta$ is less than or equal to some fixed finite constant Δ, then we shall say that the boundary function $\varphi(P)$ satisfies the three-point condition with the constant Δ.

The three-point condition obviously implies that no three distinct points P_1^*, P_2^*, P_3^* of Γ^* are on the same vertical plane (the xy-plane being thought of as horizontal). This again implies that the curve Γ, bearing the given boundary values, is convex in the strict sense. That is to say, Γ must be a convex curve no arc of which reduces to a straight segment. Obviously, however, the three-point condition requires more than that; the condition implies a restriction not only upon Γ but also upon the boundary function $\varphi(P)$ given on Γ.

If Γ is not convex, then the boundary value problem generally is not solvable (see III.17). In this sense, the convexity of Γ is a necessary restriction. On the other hand, it will follow from the results obtained for the problem of PLATEAU in the parametric form that the three-point condition can be dropped (see VI.18). In the work concerned with the non-parametric problem the three-point condition is however absolutely essential.

IV.3. The three-point condition first appears in a short announcement which HILBERT gave of his method to justify the principle of DIRICHLET[1]. The ideas sketched in that announcement have been applied by LEBESGUE, in his Thesis[2], to develop a first attack on the problem of PLATEAU in the non-parametric form.

LEBESGUE considers the following problem. Let there be given, on a convex curve Γ in the xy-plane, a continuous function $\varphi(P)$ which satisfies the three-point condition with a constant Δ. Consider all the functions $z(x, y)$ which are continuous in and on Γ, and which reduce on Γ to the given function $\varphi(P)$. Determine in this class a function $z(x, y)$ such that the area $\mathfrak{A}(z)$ of the surface $z = z(x, y)$ be a minimum. $\mathfrak{A}(z)$ denotes here the area in the sense of LEBESGUE (see I.10).

LEBESGUE proceeds as follows. Consider first functions $z(x, y)$ such that the surface $z = z(x, y)$ is a polyhedron with a given number n of vertices and such that the boundary polygon of this polyhedron is inscribed in the curve Γ^* with equation $z = \varphi(P)$, where $\varphi(P)$ is the given boundary function. For a given n, the area $\mathfrak{A}(z)$ of this polyhedron is a continuous function of a finite number of parameters, and hence, for a given n, the problem $\mathfrak{A}(z) = \text{minimum}$ is an ordinary minimum problem and it has consequently a solution $z = z_n(x, y)$. It follows then from the definition of the area, in the sense of LEBESGUE, that $\mathfrak{A}(z_n)$ converges toward the greatest lower bound m of the areas $\mathfrak{A}(z)$ of all the surfaces $z = z(x, y)$ such that $z(x, y)$ is continuous in and on Γ, and reduces to $\varphi(P)$ on Γ.

Suppose then that the sequence $z_n(x, y)$ contains a uniformly convergent subsequence $z_{n_k}(x, y)$. Denote by $z(x, y)$ the limit function. Then $z(x, y)$ is continuous in and on Γ, and reduces to $\varphi(P)$ on Γ. Hence $m \leqq \mathfrak{A}(z)$. On the other hand, on account of the lower semi-continuity of the area, $\mathfrak{A}(z) \leqq \lim \mathfrak{A}(z_{n_k}) = m$. Consequently $\mathfrak{A}(z) = m$, that is to say $z(x, y)$ solves the proposed problem $\mathfrak{A}(z) = \text{minimum}$.

The point therefore is to show that the sequence $z_n(x, y)$ contains a uniformly convergent subsequence. This will be established if it is shown that the functions $z_n(x, y)$ satisfy the LIPSCHITZ condition with the same constant, on account of a well-known theorem of ARZELÀ.

---

[1] HILBERT: Über das DIRICHLETsche Prinzip. Jber. Deutsch. Math.-Vereinig. Vol. 8 (1900) pp. 184—188.

[2] Intégrale, longueur, aire. Ann. Mat. pura appl. Vol. 7 (1920) pp. 231—359.

LEBESGUE shows then that the functions $z_n(x, y)$ satisfy the LIPSCHITZ condition with the constant Δ of the three-point condition. He reasons as follows. The polyhedron $z = z_n(x, y)$ has no convex vertices; otherwise it would be possible to lessen its area without increasing the number of its vertices, in contradiction with its minimizing property. Since the polyhedron has no convex vertices, no plane can intersect it in a closed polygon. From this it follows that every plane which contains one of the faces of the polyhedron must intersect the curve Γ^*, in which the boundary polygon is inscribed, in at least three distinct points. Hence, on account of the three-point condition, the tangent of the positive acute angle between the xy-plane and any face of the polyhedron is less than or equal to the constant Δ of the three-point condition. This obviously implies that $z_n(x, y)$ satisfies the LIPSCHITZ condition with the constant Δ.

IV.4. This reasoning is essentially the same as that sketched by HILBERT for the DIRICHLET problem in the announcement mentioned above. We do not insist on details since every step of the reasoning will be generalized and discussed in connection with later investigations. It is however important to evaluate the implications of the result of LEBESGUE for the problem of PLATEAU.

If the solution $z(x, y)$ of the problem $\mathfrak{A}(z) =$ minimum would be sufficiently regular, the area $\mathfrak{A}(z)$ of the surface $z = z(x, y)$ would be given by the classical integral

$$\iint (1 + p^2 + q^2)^{\frac{1}{2}}\, dx\, dy.$$

The function $z(x, y)$ would then be a solution, with given boundary values, of the variation problem

$$\iint (1 + p^2 + q^2)^{\frac{1}{2}}\, dx\, dy = \text{minimum}.$$

The classical methods of the Calculus of Variations would then show that $z(x, y)$ satisfies the partial differential equation

$$(1 + q^2) r - 2pqs + (1 + p^2) t = 0.$$

There remains therefore the problem of showing that the solution $z = z(x, y)$ of the problem $\mathfrak{A}(z) =$ minimum has the differential coefficients necessary for the application of the classical methods of the Calculus of Variations. This is the sense in which LEBESGUE interpreted his result as a preliminary step toward the solution of the problem of PLATEAU. We are going to review presently the investigations which resulted in the solution of the problem left open by LEBESGUE.

IV.5. The problem clearly consists of handling the solution of a two-dimensional variation problem under much more adverse conditions than those considered in the classical Calculus of Variations. In the case of one-dimensional variation problems, the well-known lemma of DU BOIS REYMOND may be considered as a step in this direction. The

4*

generalization of this lemma for two-dimensional variation problems has been obtained by HAAR[1] and will play an important part in the sequel.

Lemma of HAAR. Given, in a JORDAN region R of the xy-plane, two continuous[2] functions $u(x, y)$, $v(x, y)$ such that

$$\iint_R (u\zeta_x + v\zeta_y)\,dx\,dy = 0$$

for every function $\zeta(x, y)$ which is continuous in R, vanishes on the boundary of R, and which has continuous partial derivatives of the first order in the interior of R. Then there exists, in the interior of R, a single-valued function $\omega(x, y)$ such that

$$\omega_y = u, \quad \omega_x = -v.$$

The point to the lemma of HAAR is that no assumptions are made concerning the existence of the partial derivatives of $u(x, y)$, $v(x, y)$. Indeed, if we suppose that $u(x, y)$, $v(x, y)$ have continuous first partial derivatives, the lemma of HAAR reduces to the classical argument used in Calculus of Variations to discuss the first variation of a double integral[3].

IV.6. The original proof of HAAR has been simplified by LICHTENSTEIN[4], SCHAUDER[5] and by HAAR[6] himself. Some of the arguments used in these proofs deserve separate consideration.

a) *The one-dimensional lemma.* If $f(x)$ is continuous in a closed interval $a \leqq x \leqq b$ and is such that

$$\int_a^b f(x)\,\varphi'(x)\,dx = 0$$

for every $\varphi(x)$ which is continuous in $a \leqq x \leqq b$, vanishes for $x = a$ and $x = b$, and which has a continuous first differential coefficient in $a < x < b$, then $f(x)$ is constant.

The following proof[7] apparently is not generally known. Let $\varphi(x)$ have the properties described above and denote by γ any constant. Then

$$\int_a^b (f(x) - \gamma)\,\varphi'(x)\,dx = \int_a^b f(x)\,\varphi'(x)\,dx - \gamma \int_a^b \varphi'(x)\,dx = 0 - 0 = 0. \quad (4.1)$$

---

[1] Über die Variation der Doppelintegrale. J. reine angew. Math. Vol. 149 (1919) pp. 1—18.

[2] Later on HAAR generalized this lemma in several ways. See the expository presentation by A. HAAR: Zur Variationsrechnung. Abh. math. Semin. Hamburg. Univ. Vol. 8 (1930).

[3] See for instance BOLZA: Vorlesungen über Variationsrechnung, pp. 653—655.

[4] Bemerkungen über das Prinzip der virtuellen Verrückungen. Ann. Soc. Polon. math. (1924).

[5] Über die Umkehrung eines Satzes aus der Variationsrechnung. Acta Litt. Sci. Szeged Vol. 4 (1929) pp. 38—50.

[6] Zur Variationsrechnung. Abh. math. Semin. Hamburg. Univ. Vol. 8 (1930).

[7] See G. A. BLISS: Calculus of Variations (No. 1 of the Carus Mathematical Monographs).

Choose

$$\gamma = \frac{1}{b-a}\int_a^b f(x)\,dx.$$

Then

$$\varphi(x) = \int_a^x (f(\xi) - \gamma)\,d\xi$$

obviously has the desired properties. Since

$$\varphi'(x) = f(x) - \gamma$$

it follows from (4.1) that

$$\int_a^b (f(x) - \gamma)^2 dx = 0.$$

Hence $f(x) \equiv \gamma$.

b) *An integral test for complete differentials.* Given, in a closed rectangle $R: a \leqq x \leqq b,\ c \leqq y \leqq d$, two continuous functions $u(x, y)$, $v(x, y)$; the expression $u\,dy - v\,dx$ is called a complete differential if there exists, in the interior of R, a single-valued function $\omega(x, y)$ such that $\omega_x = -v$, $\omega_y = u$. If u, v have continuous derivatives of the first order in the interior of R, then $u\,dy - v\,dx$ is a complete differential if and only if $u_x + v_y \equiv 0$. This is the classical differential test. The vanishing of the line integral

$$\int u\,dy - v\,dx$$

for every closed, sufficiently good, curve in R is a classical integral test which remains valid under the single assumption that u, v are continuous. This is the test used by HAAR in the original proof of his lemma.

SCHAUDER[1], in his proof of the lemma of HAAR, used another interesting integral test. *Put*

$$U(x, y) = \int_c^y u(x, \eta)\,d\eta, \quad V(x, y) = \int_a^x v(\xi, y)\,d\xi. \tag{4.2}$$

Then $u\,dy - v\,dx$ is a complete differential if and only if there exist two functions $\varphi(x)$, $\psi(y)$ of the single variables x, y such that

$$U(x, y) + V(x, y) = \varphi(x) + \psi(y).$$

Indeed, suppose that $u\,dy - v\,dx$ is a complete differential: $u\,dy - v\,dx = d\omega$. Then obviously

$$U(x, y) = \omega(x, y) - \omega(x, c),$$
$$V(x, y) = \omega(a, y) - \omega(x, y),$$

and consequently

$$U(x, y) + V(x, y) = \omega(a, y) - \omega(x, c).$$

Suppose secondly that

$$U(x, y) + V(x, y) = \varphi(x) + \psi(y),$$

and put

$$\omega(x, y) = U(x, y) - \varphi(x) = -V(x, y) + \psi(y).$$

Then

$$\omega_x = -V_x = -v, \quad \omega_y = U_y = u.$$

[1] Über die Umkehrung eines Satzes aus der Variationsrechnung. Acta Litt. Sci. Szeged Vol. 4 (1929) pp. 38–50.

IV.7. The lemma of HAAR now may be proved as follows[1]. First it is clear that it is sufficient to prove the lemma for a rectangle $R: a \leqq x \leqq b, c \leqq y \leqq d$. Choose any function $l(x)$ which is continuous in $a \leqq x \leqq b$, vanishes for $x = a$ and $x = b$, and which has a continuous differential coefficient $l'(x)$ for $a < x < b$. Choose any function $m(y)$ having the same properties with respect to the interval $c \leqq y \leqq d$. Then $\zeta(x, y) = l(x) m(y)$ has all the properties required in the statement of the lemma of HAAR, and thus

$$\int_a^b \int_c^d (u l' m + v l m') \, dx \, dy = 0.$$

Partial integration gives

$$\int_a^b \int_c^d H l' m' \, dx dy = 0, \tag{4.3}$$

where $H = U + V$, and U, V are given by (4.2). It follows then from (4.3), on account of the one-dimensional lemma (see IV.6), that

$$\int_c^d H(x, y) m'(y) \, dy$$

is constant, that is to say independent of x. Hence

$$\int_c^d H(x, y) m'(y) \, dy = \int_c^d H(a, y) m'(y) \, dy,$$

or

$$\int_c^d [H(x, y) - H(a, y)] m'(y) \, dy = 0.$$

Applying again the one-dimensional lemma, we see that $H(x, y) - H(a, y)$ is constant for every fixed value of x. Hence

$$H(x, y) - H(a, y) = H(x, c) - H(a, c) = H(x, c), \tag{4.4}$$

since obviously $H(a, c) = 0$. (4.4) gives

$$H = U + V = H(a, y) + H(x, c).$$

It follows then from the integral test of SCHAUDER that $u dy - v dx$ is a complete differential, as asserted by the lemma of HAAR.

IV.8. A. HAAR applied his lemma to variation problems of the form $\iint F(p, q) \, dx dy =$ minimum in the following way[2]. Consider, in a JORDAN region R, all the functions $z(x, y)$ which are continuous in R, which reduce on the boundary of R to a given boundary function, and which have continuous partial derivatives $z_x = p$, $z_y = q$ in the interior

[1] J. SCHAUDER: Über die Umkehrung eines Satzes aus der Variationsrechnung. Acta Litt. Sci. Szeged Vol. 4 (1929) pp. 38–50. — A. HAAR: Zur Variationsrechnung. Abh. math. Semin. Hamburg. Univ. Vol. 8 (1930).

[2] A. HAAR: Über die Variation der Doppelintegrale. J. reine angew. Math. Vol. 149 (1919) pp. 1–18.

of R. Suppose there exists in this class a function $z(x, y)$ which solves the variation problem

$$\iint_R F(p, q)\,dx\,dy = \text{minimum}.$$

Let $\zeta(x, y)$ denote any function which is continuous in R, vanishes on the boundary of R, and which has continuous partial derivatives of the first order in the interior of R. Let ε be a small parameter and put

$$J(\varepsilon) = \iint_R F(p + \varepsilon\zeta_x, q + \varepsilon\zeta_y)\,dx\,dy.$$

On account of the minimizing property of $z(x, y)$, this function $J(\varepsilon)$ has a minimum for $\varepsilon = 0$. Hence

$$J'(0) = \iint_R (F_p\zeta_x + F_q\zeta_y)\,dx\,dy = 0.$$

There follows then from the lemma of HAAR the existence, in the interior of R, of a single-valued function $\omega(x, y)$ such that

$$F_p = \omega_y, \quad F_q = -\omega_x. \tag{4.5}$$

If the minimizing function $z(x, y)$ were known to have continuous partial derivatives of the second order also, then it would follow from (4.5) that

$$\frac{\partial}{\partial x}F_p + \frac{\partial}{\partial y}F_q = F_{pp}r + 2F_{pq}s + F_{qq}t = 0. \tag{4.6}$$

This is the classical EULER-LAGRANGE equation of the problem. Examples show however that the minimizing function need not have partial derivatives of the second order. The lemma of HAAR permits us to derive the differential equations (4.5) under the single assumption of the existence and continuity of the partial derivatives of the first order.

IV.9. A. HAAR proposed the problem of applying the equations (4.5) to the study of the analytic character of the solutions of positive regular analytic variation problems of the form $\iint F(p, q)\,dx\,dy = \text{minimum}$. If $F(p, q)$ is an analytic function of p, q and if it satisfies the inequalities

$$F_{pp} > 0, \quad F_{qq} > 0, \quad F_{pp}F_{qq} - F_{pq}^2 > 0 \tag{4.7}$$

for all values of p, q, then the problem $\iint F(p, q)\,dx\,dy = \text{minimum}$ is called positive regular and analytic. If $z(x, y)$ is a solution, with continuous partial derivatives of the second order, of such a problem, then $z(x, y)$ satisfies the EULER-LAGRANGE equation (4.6). This is a quasi-linear equation of the second order, and it is of the elliptic type, on account of (4.7). Every solution of such an equation is analytic as soon as it has continuous partial derivatives of the first and second order[1]. Hence every solution of a regular analytic variation problem $\iint F(p, q)\,dx\,dy = \text{minimum}$ is analytic as soon as it has continuous

[1] See L. LICHTENSTEIN: Neuere Entwicklung usw. Enzyklopädie der math. Wiss. Vol. 2 (3) pp. 1277—1334.

partial derivatives of the first and second order. The problem proposed by HAAR requires the obtaining of this conclusion under the assumption of the existence and continuity of the first partial derivatives only.

At the present time the solution of this problem is known only in two special cases, namely for the DIRICHLET problem

$$\tfrac{1}{2}\iint (p^2 + q^2)\, dx\, dy = \text{minimum},$$

and for the area-problem

$$\iint (1 + p^2 + q^2)^{\frac{1}{2}}\, dx\, dy = \text{minimum}.$$

In the case of the DIRICHLET problem, the equations of HAAR reduce to

$$p = z_x = \omega_y, \quad q = z_y = -\omega_x,$$

that is to say to the CAUCHY-RIEMANN equations. The fact that $z(x, y)$ is analytic follows therefore from classical theorems in the theory of functions of a complex variable.

We are going to review now the case $F = (1 + p^2 + q^2)^{\frac{1}{2}}$.

IV.10. We consider first the area problem in the parametric form[1]. Let $S: \mathfrak{x} = \mathfrak{x}(u, v)$, $u^2 + v^2 \leqq 1$, be a regular surface of class C' (see II.2) bounded by a given curve, and suppose that the area of S is a minimum.

Consider an interior point P_0 of S. Choose the coordinate system xyz in such a way that none of the coordinate planes is parallel to the normal of S at P_0. Denote by S_0 a small simply connected portion of S comprising P_0. If S_0 is sufficiently small, then S_0 can be represented in any one of the three forms $z = z(x, y)$, $x = x(y, z)$, $y = y(z, x)$, where $z(x, y), x(y, z), y(z, x)$ are single-valued functions with continuous first partial derivatives.

Denote by X, Y, Z the direction-cosines of the normal to the surface. The first step is to prove that the three expressions

$$Y dz - Z dy, \quad Z dx - X dz, \quad X dy - Y dx$$

are complete differentials on S_0.[2]

Consider the representation $z = z(x, y)$ of S_0. Then $z(x, y)$ can be considered as a solution of the variation problem

$$\iint (1 + p^2 + q^2)^{\frac{1}{2}}\, dx\, dy = \text{minimum}.$$

Since $z(x, y)$ has continuous first partial derivatives, the result of HAAR can be applied. The equations of HAAR reduce in the present case to

$$\frac{p}{(1 + p^2 + q^2)^{\frac{1}{2}}} = \omega_y, \quad \frac{q}{(1 + p^2 + q^2)^{\frac{1}{2}}} = -\omega_x.$$

[1] T. RADÓ: Über den analytischen Charakter der Minimalflächen. Math. Z. Vol. 24 (1925) pp. 321–327.

[2] Cf. II.11.

In other words, the expression

$$\frac{p\,dy - q\,dx}{(1+p^2+q^2)^{\frac{1}{2}}}$$

is a complete differential[1]. On the other hand, we have

$$X = -\frac{p}{(1+p^2+q^2)^{\frac{1}{2}}}, \quad Y = -\frac{q}{(1+p^2+q^2)^{\frac{1}{2}}},$$

and consequently

$$X\,dy - Y\,dx = -\frac{p\,dy - q\,dx}{(1+p^2+q^2)^{\frac{1}{2}}}.$$

This proves that $X\,dy - Y\,dx$ is a complete differential. Using the representations $x = x(y, z)$, $y = y(z, x)$ of S_0, we see that $Y\,dz - Z\,dy$, $Z\,dx - X\,dz$ are also complete differentials.

Consider then again the representation $z = z(x, y)$ of S_0. It can be supposed that the xy-projection of S_0 is a rectangle $R: x_1 \leqq x \leqq x_2$, $y_1 \leqq y \leqq y_2$. We have for the components X, Y, Z of the unit normal vector

$$X = -\frac{p}{W}, \quad Y = -\frac{q}{W}, \quad Z = \frac{1}{W}, \quad W = (1+p^2+q^2)^{\frac{1}{2}}.$$

The preceding result is then expressed by the equations

$$\left.\begin{aligned} &-\frac{pq}{W}\,dx - \frac{1+q^2}{W}\,dy = d\omega_1, \quad \frac{1+p^2}{W}\,dx + \frac{pq}{W}\,dy = d\omega_2, \\ &\frac{q}{W}\,dx - \frac{p}{W}\,dy = d\omega_3, \end{aligned}\right\} \tag{4.8}$$

where $\omega_1, \omega_2, \omega_3$ denote three auxiliary functions which are single-valued in R. By assumption p, q and consequently the first partial derivatives of $\omega_1, \omega_2, \omega_3$ are continuous. Introduce new parameters α, β by the equations

$$\alpha = x, \quad \beta = \omega_1(x, y).$$

From $\beta_y = \omega_{1y} = -\frac{1+q^2}{W} < 0$ it follows then that R is carried in a one-to-one way into a certain region R^* of the $\alpha\beta$-plane. Direct computation shows then (cf. II.16) that x and ω_1, y and ω_2, z and ω_3, as functions of α and β, satisfy in R^* the CAUCHY-RIEMANN equations

$$x_\alpha = \omega_{1\beta}, \quad x_\beta = -\omega_{1\alpha},$$
$$y_\alpha = \omega_{2\beta}, \quad y_\beta = -\omega_{2\alpha},$$
$$z_\alpha = \omega_{3\beta}, \quad z_\beta = -\omega_{3\alpha}.$$

Since $x, y, z, \omega_1, \omega_2, \omega_3$ as functions of α, β have continuous first partial derivatives, it follows from classical theorems that the six functions involved are analytic functions of α and β. Since

$$\frac{\partial(\alpha, \beta)}{\partial(x, y)} = -\frac{1+q^2}{W} \neq 0,$$

[1] Cf. II.6.

it follows from well-known theorems on implicit functions that α, β are analytic functions of x and y. As z is an analytic function of α, β and as α, β are analytic functions of x and y, it follows that z is an analytic function of x and y. Thus we have the theorem[1]:

If a regular surface S of class C', bounded by a given curve, has a minimum area, then S is analytic (and consequently is a minimal surface).

IV.11. The minimizing property of S has been used only to establish the equations (4.8); from that point on, only those equations were used. Hence we have the following theorem[2].

Let $z(x, y)$ be a function which has continuous partial derivatives of the first order in a JORDAN region R. Suppose there exist in R three single-valued functions $\omega_1, \omega_2, \omega_3$ such that

$$\omega_{1x} = -\frac{pq}{W}, \quad \omega_{1y} = -\frac{1+q^2}{W},$$

$$\omega_{2x} = \frac{1+p^2}{W}, \quad \omega_{2y} = \frac{pq}{W},$$

$$\omega_{3x} = \frac{q}{W}, \quad \omega_{3y} = -\frac{p}{W}.$$

Then $z(x, y)$ is an analytic function of x, y and satisfies the partial differential equation

$$(1 + q^2)r - 2pqs + (1 + p^2)t = 0.$$

The second half of the statement follows of course from the first half directly by differentiation.

IV.12. Consider now *the area problem in the non-parametric form*

$$\iint (1 + p^2 + q^2)^{\frac{1}{2}} dx\, dy = \text{minimum}.$$

Suppose we have, in a JORDAN region R, a solution $z(x, y)$ with given boundary values and having continuous first partial derivatives in the interior of R. Using variations of the form $z + \varepsilon\zeta$, we obtain from the lemma of HAAR the result that $X dy - Y dx$ is a complete differential. The surface $z = z(x, y)$ is now only known to have a smallest area with respect to surfaces of the same form. The reasoning which showed, in the case of the parametric problem, that $Y dz - Z dy$, $Z dx - X dz$ are also complete differentials is therefore not legitimate in the present case. Still, the result remains valid[2]. The situation can be handled by the so-called *method of the variation of the independent variables.*

[1] T. RADÓ: Über den analytischen Charakter der Minimalflächen. Math. Z. Vol. 24 (1925) pp. 321–327.

[2] T. RADÓ: Bemerkung über die Differentialgleichungen zweidimensionaler Variationsprobleme. Acta Litt. Sci. Szeged Vol. 3 (1925) pp. 147–156.

Let us consider the general problem

$$\iint F(p, q)\, dx\, dy = \text{minimum}.$$

Let $z(x, y)$ be a solution with given boundary values, and suppose only that $z(x, y)$ has continuous first partial derivatives. It clearly is legitimate to suppose that the region R in which the problem is considered is a circle K. Denote by $t(x, y)$ a function continuous in K, vanishing on the boundary of K, and having continuous and bounded first partial derivatives in K. Let ε be a small parameter and define a transformation $T(\varepsilon)$ by the formulas

$$T(\varepsilon): x = u + \varepsilon t(u, v), \quad y = v, \quad (u, v) \text{ in } K.$$

For sufficiently small values of ε this is a one-to-one and continuous transformation of K into itself. Hence, u, v can be expressed as single-valued functions of x, y and ε. It follows in this way that the equations

$$x = u + \varepsilon t(u, v), \quad y = v, \quad z = z(u, v), \quad (u, v) \text{ in } K,$$

where z is the minimizing function under investigation, define a family of admissible surfaces. For these surfaces the integral $\iint F(p, q)\, dx\, dy$ becomes a function $J(\varepsilon)$ which has a minimum for $\varepsilon = 0$. Hence $J'(0) = 0$. After suitable transformations, the function $J(\varepsilon)$ appears in the form

$$J(\varepsilon) = \iint\limits_K F\left(\frac{p}{1 + \varepsilon t_x}, \; q - \frac{\varepsilon p t_y}{1 + \varepsilon t_x}\right)(1 + \varepsilon t_x)\, dx\, dy.$$

The condition $J'(0) = 0$ gives therefore the equation

$$\iint\limits_K [(F - pF_p) t_x - pF_q t_y]\, dx\, dy = 0.$$

It follows then from the lemma of HAAR that

$$pF_q\, dx + (F - pF_p)\, dy$$

is a complete differential. Using in a similar way variations of the independent variable y, we obtain the result that

$$(F - qF_q)\, dx + qF_p\, dy$$

is a complete differential. Finally it follows by means of the usual variation $z + \varepsilon\zeta$ that $F_q dx - F_p dy$ is a complete differential. That is to say:

If $z(x, y)$ is a solution with continuous first partial derivatives of the boundary value problem for the variation problem

$$\iint F(p, q)\, dx\, dy = \text{minimum},$$

then the three expressions

$$pF_q\, dx + (F - pF_p)\, dy, \quad (F - qF_q)\, dx + qF_p\, dy, \quad F_q dx - F_p dy$$

are complete differentials. In other words: there exist three single-valued auxiliary functions $\omega_1(x, y)$, $\omega_2(x, y)$, $\omega_3(x, y)$ which satisfy, together with the minimizing function $z(x, y)$, the following system of partial differential equations[1],

$$\left.\begin{aligned} \omega_{1x} &= pF_q, & \omega_{1y} &= F - pF_p, \\ \omega_{2x} &= F - qF_q, & \omega_{2y} &= qF_p, \\ \omega_{3x} &= F_q, & \omega_{3y} &= -F_p. \end{aligned}\right\} \tag{4.9}$$

The last two equations are those obtained by HAAR himself. In case the minimizing function $z(x, y)$ is known to have continuous partial derivatives of the second order, the auxiliary functions $\omega_1, \omega_2, \omega_3$ can be eliminated by crosswise differentiation and there results the classical EULER-LAGRANGE equation. Conversely, it is immediate that if $z(x, y)$ satisfies the EULER-LAGRANGE equation, then the auxiliary functions $\omega_1, \omega_2, \omega_3$ exist. Indeed, the EULER-LAGRANGE equation is the common integrability condition for the three couples of equations comprised in the above system. The point is again that the equations (4.9) can be established under the single assumption that the minimizing function has continuous partial derivatives of the first order.

IV.13. In the case $F = (1 + p^2 + q^2)^{\frac{1}{2}}$ the equations (4.9) reduce to those considered in IV.11. Hence, on account of the remark made there, we have the following theorem[1].

If $z(x, y)$ is a solution, with continuous partial derivatives of the first order, of the boundary value problem for the variation problem

$$\iint (1 + p^2 + q^2)^{\frac{1}{2}} dx\,dy,$$

then $z(x, y)$ is analytic and satisfies the partial differential equation

$$(1 + q^2)r - 2pqs + (1 + p^2)t = 0. \tag{4.10}$$

IV.14. We are going to review at present a paper of A. HAAR[2] in which he gave important generalizations and applications of the results discussed so far in this Chapter. In § 2 and § 3 of his paper, HAAR generalizes the theorem of IV.13 for the case when the minimizing function $z(x, y)$ is only known to satisfy the LIPSCHITZ condition. If a function $z(x, y)$ satisfies the condition of LIPSCHITZ in a domain D, then $p = z_x$ and $q = z_y$ exist almost everywhere in D and are bounded and measurable functions[3]. The expression $(1 + p^2 + q^2)^{\frac{1}{2}}$ is then also a bounded and measurable function of x, y and hence the integral

$$\iint_D (1 + p^2 + q^2)^{\frac{1}{2}} dx\,dy$$

[1] T. RADÓ: Bemerkung über die Differentialgleichungen zweidimensionaler Variationsprobleme. Acta Litt. Sci. Szeged Vol. 3 (1925) pp. 147—156.

[2] A. HAAR: Über das PLATEAUsche Problem. Math. Ann. Vol. 97 (1927) pp. 124—258.

[3] The theory of functions of two variables, satisfying the LIPSCHITZ condition, has been the object of important investigations of H. RADEMACHER: Über partielle und totale Differenzierbarkeit I, II. Math. Ann. Vol. 79 (1919) pp. 340—359 and Vol. 81 (1920) pp. 52—63.

exists in the LEBESGUE sense. Thus there is a good sense to the statement of the variation problem

$$\iint_D (1+p^2+q^2)^{\frac{1}{2}}\,dx\,dy = \text{minimum} \tag{4.11}$$

for the class of functions $z(x, y)$ which satisfy the LIPSCHITZ condition. Using the modern theory of functions of real variables, HAAR verifies that all the arguments which have been used to prove the theorem of IV.13 remain valid under the single assumption that the minimizing function $z(x, y)$ satisfies the LIPSCHITZ condition. He obtains in this way the theorem:

If $z(x, y)$ is a solution, satisfying the condition of LIPSCHITZ, *of the boundary value problem for the variation problem* (4.11), *then $z(x, y)$ is analytic and satisfies the partial differential equation* (4.10).

IV.15. HAAR observes that this result can be applied to the existence theorem obtained by LEBESGUE (see IV.3). Indeed, GEÖCZE proved (see I.13) that if $z(x, y)$ satisfies the condition of LIPSCHITZ, then the area $\mathfrak{A}(z)$ of the surface $z = z(x, y)$ is given by the classical integral

$$\iint (1+p^2+q^2)^{\frac{1}{2}}\,dx\,dy. \tag{4.12}$$

Hence the result of LEBESGUE concerning the problem $\mathfrak{A}(z) = \text{minimum}$ (see IV.3) implies the following existence theorem.

Let there be given, on a convex JORDAN *curve Γ in the xy-plane, a function $\varphi(P)$ of the point P varying on Γ which satisfies the three-point condition with some constant Δ. Consider all the functions $z(x, y)$ which satisfy, in the* JORDAN *region bounded by Γ, the* LIPSCHITZ *condition and which reduce on Γ to the given function $\varphi(P)$. Then there exists in this class a function $z_0(x, y)$ which minimizes the integral* (4.12) *extended over the interior of Γ.*

IV.16. It follows then from the theorem in IV.14 that this function $z_0(x, y)$ is analytic and satisfies the partial differential equation (4.10). Hence the theorem:

Let there be given, on a convex JORDAN *curve Γ, a function $\varphi(P)$ which satisfies the three-point condition with some constant Δ. Then there exists, in the* JORDAN *region bounded by Γ, a solution of the partial differential equation*

$$(1+q^2)r - 2pqs + (1+p^2)t = 0,$$

which reduces to $\varphi(P)$ on Γ.

IV.17. In § 1 of his paper under discussion, HAAR generalizes the existence theorem of IV.15 for positive regular variation problems of the form

$$\iint F(p, q)\,dx\,dy = \text{minimum}.$$

Such a problem is called positive regular if F satisfies the inequalities

$$F_{pp} > 0, \quad F_{qq} > 0, \quad F_{pp}F_{qq} - F_{pq}^2 > 0 \tag{4.13}$$

for all values of p and q. The existence proof presented by HAAR for this general case is considerably simpler than the existence proof which follows from the results of LEBESGUE and GEÖCZE for the special case $F = (1 + p^2 + q^2)^{\frac{1}{2}}$. The reviewer believes that this justifies the detailed discussion of the general case in this report. We shall first consider the main arguments used by HAAR and we shall then describe the existence proof itself.

IV.18. Let $z(x, y)$ be a function continuous in a JORDAN region bounded by a convex JORDAN curve Γ. Suppose that the boundary values of $z(x, y)$ satisfy the three-point condition with some constant Δ. Suppose also that $z(x, y)$ has the property that there exists no plane which intersects the surface $z = z(x, y)$ in a closed curve (this assumption will be stated in a more exact form presently). One of the main arguments of HAAR is then the theorem that under these circumstances the function $z(x, y)$ satisfies the LIPSCHITZ condition with the constant Δ of the three-point condition.

Similar arguments, concerned with various special types of surfaces $z = z(x, y)$, played an important part in most of the investigations on the problem of PLATEAU in the non-parametric form[1]. For instance, the method of LEBESGUE reviewed in IV.3 is based on the special case concerned with polyhedrons. The general theorem has been stated by HAAR in the course of preliminary investigations on the problem of PLATEAU and has been proved first by T. RADÓ[2]. A much simpler proof has then been given by J. v. NEUMANN[3].

The exact statement of the theorem is based on the following definitions. Let $f(x, y)$ be continuous in a JORDAN region R. Let D be any domain (connected open set) comprised in R. Denote by D^* the set of all boundary points of D, and by M^*, m^* the maximum and minimum respectively of $f(x, y)$ on D^*. The function $f(x, y)$ is called *monotonic*[4] in the JORDAN region R if the condition

$$m^* \leqq f(x, y) \leqq M^* \text{ for } (x, y) \text{ in } D$$

is satisfied for every domain D in R.

Let a, b, c be three constants. If the function $f(x, y) - (ax + by + c)$ is monotonic in R for every choice of the constants a, b, c, then $f(x, y)$ is called a *saddle-function* in R. Using this terminology, we may state the theorem as follows.

[1] See for references A. HAAR: Über das PLATEAUsche Problem. Math. Ann. Vol. 97 (1927) p. 127 and 141.

[2] T. RADÓ: Geometrische Betrachtungen über zweidimensionale reguläre Variationsprobleme. Acta Litt. Sci. Szeged Vol. 2 (1926) pp. 228—253.

[3] J. v. NEUMANN: Über einen Satz der Variationsrechnung. Abh. math. Sem. Hamburg. Univ. Vol. 8 (1931) pp. 28—31.

[4] H. LEBESGUE: Intégrale, longueur, aire. Ann. Mat. pura appl. Vol. 7 (1902) pp. 231—359.

If $z(x, y)$ is a saddle-function in a convex JORDAN *region R, and if the boundary values of $z(x, y)$ satisfy the three-point condition with some constant Δ, then $z(x, y)$ satisfies in R the* LIPSCHITZ *condition with the constant Δ.*

IV.19. The proof of this theorem, although quite elementary, is rather involved even in the simplified form due to J. v. NEUMANN, and therefore we restrict ourselves to the following remarks. Let (x_1, y_1), (x_2, y_2) be any two distinct points in R; the proof consists of showing that the inequality

$$\frac{|z(x_2, y_2) - z(x_1, y_1)|}{[(x_2 - x_1)^2 + (y_2 - y_1)^2]^{\frac{1}{2}}} > \Delta \tag{4.14}$$

is impossible. This is practically obvious if at least one of the two points is on the boundary of R[1]. The complications arise solely in the general case when both points are interior points of R.

T. RADÓ observed that as far as the existence proof of HAAR is concerned, these complications easily can be avoided[1]. Indeed, the functions $z(x, y)$, to which the theorem will be applied in the course of the existence proof, will possess the following additional property. Let h, k be two constants. Denote by $R^{h,k}$ the JORDAN region obtained from the JORDAN region R by the transformation $\bar{x} = x + h, \bar{y} = y + k$. Then the function $z(x - h, y - k) - z(x, y)$ is monotonic in the region consisting of the common points of R and $R^{h,k}$, for every choice of the constants h and k.

Consider then two interior points (x_1, y_1), (x_2, y_2) of R such that (4.14) is satisfied. Put $h = x_2 - x_1$, $k = y_2 - y_1$. From (4.14) and from the monotonic character of $z(x - h, y - k) - z(x, y)$ it follows then immediately that (4.14) is satisfied also for a couple of points $(\bar{x}_1, \bar{y}_1)$, $(\bar{x}_2, \bar{y}_2)$, at least one of which is a boundary point of R. Hence it is sufficient to disprove (4.14) in the almost trivial case when at least one of the two points is on the boundary[1].

IV.20. Another important argument in the existence proof of HAAR is the following theorem.

Suppose that $F(p, q)$ satisfies, for all values of p and q, the inequalities (4.13). *Consider then, in a* JORDAN *region R, a sequence $z_n(x, y)$ converging uniformly to a function $z(x, y)$. Suppose that all these functions satisfy in R the* LIPSCHITZ *condition (not necessarily with the same constant). Then*

$$\varliminf \iint_R F(p_n, q_n)\, dx\, dy \geqq \iint_R F(p, q)\, dx\, dy,$$

where $p_n = z_{nx}$, $q_n = z_{ny}$, $p = z_x$, $q = z_y$.

The theorem expresses the *lower semi-continuity* of the integral $\iint F(p, q)\, dx\, dy$. In the special case $F = (1 + p^2 + q^2)^{\frac{1}{2}}$, the integral

[1] T. RADÓ: Über zweidimensionale reguläre Variationsprobleme. Math. Ann. Vol. 101 (1929) pp. 620—632. See in particular § 1, No. 2.

is equal to the area of the surface $z = z(x, y)$ in the sense of LEBESGUE (see I.13), and hence the theorem is a consequence of the lower semi-continuity of the area (see I.12). The lower semi-continuity of the area-integral has been generalized to various classes of simple and multiple integrals[1].

For the integral in which we are interested, a very simple proof is obtained by applying a device used by HAAR[2] to the method which results from the work of GEÖCZE for the special case $(1 + p^2 + q^2)^{\frac{1}{2}}$. This proof proceeds in the following steps[3]. First observe that it is sufficient to consider the case when the JORDAN region R is a square. Use the notation

$$J(z) = \iint\limits_R F(p, q)\, dx\, dy,$$

where z is any function which satisfies the condition of LIPSCHITZ in R. Subdivide R into m^2 congruent small squares, and put

$$\Sigma_m(z) = \sum_{k=1}^{m^2} F\left(\frac{1}{\sigma_{m,k}} \iint\limits_{\sigma_{m,k}} p\, dx\, dy,\ \frac{1}{\sigma_{m,k}} \iint\limits_{\sigma_{m,k}} q\, dx\, dy\right) \sigma_{m,k}.$$

In this formula, $\sigma_{m,1}, \sigma_{m,2}, \ldots, \sigma_{m,m^2}$ denote the small squares into which R has been subdivided; $\sigma_{m,k}$ also denotes the area of $\sigma_{m,k}$. The following properties of $\Sigma_m(z)$ will be used.

1. $\Sigma_m(z) \to J(z)$ for $m \to \infty$.
2. $\Sigma_m(z) \leqq J(z)$.
3. If a sequence $z_n(x, y)$ converges uniformly to a function $z(x, y)$ in R, and if all these functions satisfy the condition of LIPSCHITZ in R (not necessarily with the same constant), then

$$\lim_{n\to\infty} \Sigma_m(z_n) = \Sigma_m(z)$$

for every fixed value of m.

The proofs of 1 and 3 are immediate if all the functions concerned are supposed to have continuous first partial derivatives. The validity of the arguments used in this elementary case can then be extended on account of well-known general theorems of the theory of functions of real variables. As to 2, observe first that we have, on account of the inequalities (4.13),

$$F\left(\frac{a_1 + \cdots + a_\nu}{\nu}, \frac{b_1 + \cdots + b_\nu}{\nu}\right) \leqq \frac{F(a_1, b_1) + \cdots + F(a_\nu, b_\nu)}{\nu}$$

[1] See, also for references, L. TONELLI: Sur la semi-continuité des intégrales doubles du Calcul des Variations. Acta math. Vol. 53 (1929) pp. 325–346 and J. E. MCSHANE: On the semi-continuity of double integrals. Ann. of Math. Vol. 33 (1932) pp. 460–484.

[2] Über das PLATEAUsche Problem. Math. Ann. Vol. 97 (1927) pp. 124–258.

[3] T. RADÓ: Über zweidimensionale reguläre Variationsprobleme. Math. Ann. Vol. 101 (1929) pp. 620–632.

for every value of the positive integer ν and of the real constants $a_1, \ldots, a_\nu, b_1, \ldots, b_\nu$.[1] The inequality 2 follows then immediately by a passage to the limit. The main theorem

$$\underline{\lim}\, J(z_n) \geqq J(z)$$

is then proved as follows. 2 gives $\Sigma_m(z_n) \leqq J(z_n)$. For $n \to \infty$ it follows, on account of 3, that $\Sigma_m(z) \leqq \underline{\lim}\, J(z_n)$. For $m \to \infty$, it follows, on account of 1, that $J(z) \leqq \underline{\lim}\, J(z_n)$.

IV.21. The statement of the existence theorem of HAAR is as follows.

Let there be given, on a convex curve Γ of the xy-plane, a continuous function $\varphi(P)$ of the point P varying on Γ which satisfies the three-point condition with some constant Δ. Consider the class of all functions $z(x, y)$ which satisfy the LIPSCHITZ *condition in the* JORDAN *region R bounded by Γ and reduce to $\varphi(P)$ on Γ. Then there exists in this class a solution of the variation problem*

$$\iint_R F(p, q)\, dx\, dy = \text{minimum}, \tag{4.15}$$

if the problem is positive regular, that is to say if

$$F_{pp} > 0, \quad F_{qq} > 0, \quad F_{pp}F_{qq} - F_{pq}^2 > 0$$

for all values of p and q.

IV.22. The proof proceeds in the following steps. Denote by K the class of functions described above. Denote by K_L the class of those functions in K which satisfy the LIPSCHITZ condition with a given constant L. A first remark is then that for $L \geqq \Delta$ the class K_L certainly is not empty. HAAR observes that the harmonic function which reduces to $\varphi(P)$ on Γ satisfies the LIPSCHITZ condition with the constant Δ and consequently is comprised in K_L for $L \geqq \Delta$. A more elementary argument is obtained by the following construction. Denote by Γ^* the curve in the xyz-space with the equation $z = \varphi(P)$. Take a fixed point P_0^* on Γ^*. Then the straight segments connecting P_0^* with a variable point P^* of Γ^* constitute a surface S with the obvious property: if A^*, B^* are any two points of S, then there exists a plane through A^*, B^* which intersects Γ^* in at least three distinct points. If then $z = z(x, y)$ is the equation of S, it follows from the three-point condition that $z(x, y)$ satisfies the LIPSCHITZ condition with the constant Δ.

IV.23. Let L be any constant $> \Delta$. Consider the problem (4.15) for the class K_L only. Then it readily is seen that this restricted problem has a solution. First, the class K_L is not empty (see IV.22). Secondly, since $|p|$, $|q|$ are both $\leqq L$, and since $F(p, q)$ is continuous, the absolute value of the integral is uniformly bounded for all functions of

[1] See for instance PÓLYA-SZEGÖ: Aufgaben und Lehrsätze. Vol. 1 pp. 51–52, problems 70 and 71.

the class K_L. The greatest lower bound m_L of the integral, for all functions of the class K_L, is therefore finite. There exists, by definition, a sequence of functions $z_n(x, y)$, comprised in K_L, such that

$$\iint_R F(p_n, q_n)\,dx\,dy \to m_L.$$

Since the functions $z_n(x, y)$ satisfy the LIPSCHITZ condition with the same constant L, the sequence $z_n(x, y)$ contains a uniformly convergent subsequence $z_{n_k}(x, y)$. The limit function $z_0(x, y)$ clearly belongs again to K_L, and thus

$$\iint_R F(p_0, q_0)\,dx\,dy \geqq m_L.$$

On the other hand, on account of IV.20,

$$\iint_R F(p_0, q_0)\,dx\,dy \leqq \underline{\lim} \iint_R F(p_{n_k}, q_{n_k})\,dx\,dy = m_L.$$

Hence

$$\iint_R F(p_0, q_0)\,dx\,dy = m_L.$$

IV.24. The function $z_0(x, y)$, obtained in this way, satisfies the LIPSCHITZ condition with the constant L, since it is comprised in the class K_L. *The heart of the existence proof is the fact that $z_0(x, y)$ satisfies the* LIPSCHITZ *condition with the constant $\Delta < L$.*

On account of IV.18, it is sufficient to verify that $z_0(x, y)$ is a saddle-function. In the special case $F = (1 + p^2 + q^2)^{\frac{1}{2}}$, the fact is geometrically obvious. Indeed, if $z_0(x, y)$ is not a saddle-function, then we have a plane $z = ax + by + c$, such that the difference $z_0(x, y) - (ax + by + c)$ vanishes on the boundary of a certain domain D, while it is, say, positive in D itself. Replace then $z_0(x, y)$ in D by $ax + by + c$. There results a new function $\bar{z}_0(x, y)$. No secant of the new surface $z = \bar{z}_0(x, y)$ is steeper than the corresponding secant of the old surface $z = z_0(x, y)$, while the area of the new surface clearly is less than the area of the old surface, in contradiction with the minimizing property of $z_0(x, y)$. The analytic justification of this geometric argument can be extended immediately to the problem with a general $F(p, q)$.

If one desires to take advantage of the remark in IV.19, then it is necessary to verify that $z_0(x, y)$ possesses the additional property stated there. The case $F = (1 + p^2 + q^2)^{\frac{1}{2}}$ permits again of obvious geometric justification, which also suggests the proof for the general case[1].

IV.25. Let now $\bar{z}(x, y)$ be any function of the class K, and let $z_0(x, y)$ be the minimizing function for the class K_L. Consider the function

$$z(x, y) = z_0(x, y) + \Theta(\bar{z}(x, y) - z_0(x, y)),$$

[1] T. RADÓ: Über zweidimensionale reguläre Variationsprobleme. Math. Ann. Vol. 101 (1929) pp. 620–632.

where Θ is a real parameter. For every given value of Θ, $z(x, y)$ clearly belongs to the class K. The integral

$$\iint_R F(p, q)\,dx\,dy$$

becomes a function $J(\Theta)$ of Θ, and it follows from the inequalities (4.13) immediately that $J''(\Theta) \geqq 0$. Hence $J(\Theta)$ is convex.

For small values of Θ, $z(x, y)$ satisfies the LIPSCHITZ condition with practically the same constant as $z_0(x, y)$. As $z_0(x, y)$ satisfies the LIPSCHITZ condition with the constant $\Delta < L$ (see IV.24), it follows that for small values of Θ the function $z(x, y)$ belongs to the class K_L. Hence, on account of the minimizing property of $z_0(x, y)$, $J(0) \leqq J(\Theta)$ for small values of Θ. A local minimum of a convex function is however necessarily an absolute minimum. Hence $J(0) \leqq J(1)$, that is to say

$$\iint_R F(p_0, q_0)\,dx\,dy \leqq \iint_R F(\bar{p}, \bar{q})\,dx\,dy. \tag{4.16}$$

In other words: $z_0(x, y)$ minimizes the integral not only in the restricted class K_L, but also in the whole class K. This proves the existence theorem stated in IV.21.

IV.26. The reasoning used in IV.25 may be replaced by the following argument. Let $\bar{z}(x, y)$ satisfy the condition of LIPSCHITZ with some constant $\bar{L}$ (since $\bar{z}$ belongs to K, such a constant exists by assumption). If $\bar{L} \leqq L$, then $\bar{z}$ belongs to K_L, and (4.16) is a direct consequence of the minimizing property of z_0. If $\bar{L} > L > \Delta$, then consider the class $K_{\bar{L}}$. Let $\bar{z}_0$ be the minimizing function for the class $K_{\bar{L}}$ ($\bar{z}_0$ exists for the same reason as z_0). Then, by the definition of $\bar{z}_0$,

$$\iint_R F(\bar{p}_0, \bar{q}_0)\,dx\,dy \leqq \iint_R F(\bar{p}, \bar{q})\,dx\,dy. \tag{4.17}$$

But $\bar{z}_0$ satisfies the condition of LIPSCHITZ with the constant Δ (for the same reason as z_0). Hence $\bar{z}_0$ belongs to the class K_L, since $\Delta < L$. Consequently, on account of the minimizing property of z_0,

$$\iint_R F(p_0, q_0)\,dx\,dy \leqq \iint_R F(\bar{p}_0, \bar{q}_0)\,dx\,dy. \tag{4.18}$$

(4.16) follows now from (4.17) and (4.18).

IV.27. Suppose $F(p, q)$ is an *analytic* function of p, q. The question naturally arises as to whether the solution $z(x, y)$ of the variation problem (4.15), obtained by HAAR, is analytic. From the existence proof it only follows that $z(x, y)$ satisfies the LIPSCHITZ condition. In the special case $F = (1 + p^2 + q^2)^{\frac{1}{2}}$, the proof of the analytic character of $z(x, y)$ depended essentially on computations suggested by differential geometry. HAAR gave a discussion of the extremals for the general case $F(p, q)$ with the purpose to generalize the main theorems of the differential geometry of minimal surfaces[1]. While the analogies are striking, his results did not permit as yet to establish the analytic character of the solution of the variation problem under the only assumption that the solution satisfies the LIPSCHITZ condition. Under the assumption that the solution has continuous first partial derivatives which satisfy the LIPSCHITZ-HÖLDER condition with some exponent λ such that $0 < \lambda < 1$, the analytic character has been proved by E. HOPF[2].

IV.28. The boundary value problem for the equation

$$(1 + q^2)r - 2pqs + (1 + p^2)t = 0$$

has been also treated in papers of A. KORN[3], CH. MÜNTZ[4], S. BERNSTEIN[5] by methods whose discussion is beyond the scope of this report. The geometrical results obtained by these authors are less general than those explicitly considered in the present review.

Chapter V.

The problem of PLATEAU in the parametric form.

V.1. The method of GARNIER[6], which we shall consider first, is concerned with problem P_3 (see III.5), and permits us to carry out a program outlined by WEIERSTRASS and then, with more details, by DARBOUX[7]. Consider a minimal surface S given by the formulas of WEIERSTRASS:

$$\left.\begin{aligned} x &= \Re\int^w (\Phi^2 - \Psi^2)\,dw, \\ y &= \Re\int^w i(\Phi^2 + \Psi^2)\,dw, \\ z &= \Re\int^w 2\Phi\Psi dw, \end{aligned}\right\} \qquad (5.1)$$

[1] A. HAAR: Über adjungierte Variationsprobleme und adjungierte Extremalflächen. Math. Ann. Vol. 100 (1928) pp. 481—502.

[2] Zum analytischen Charakter usw. Math. Z. Vol. 30 (1929) pp. 404 to 413.

[3] Über Minimalflächen, deren Randkurven wenig von ebenen Kurven abweichen. Abh. preuß. Akad. Wiss. 1909.

[4] Die Lösung des PLATEAUschen Problems über konvexen Bereichen. Math. Ann. Vol. 94 (1925) pp. 53—96.

[5] See, also for references to his previous work, Sur l'intégration des équations aux dérivées partielles du type elliptique. Math. Ann. Vol. 96 (1927) pp. 633—647.

[6] Sur le problème de PLATEAU. Ann. École norm. Vol. 45 (1928) pp. 53—144.

[7] Théorie générale des surfaces Vol. 1 Chapter 13.

and suppose that $w = u + iv$ varies in the upper half-plane $v > 0$. Reflect S upon some plane p. The resulting minimal surface S^* corresponds then to a couple of new functions Φ^*, Ψ^*, and we have the simple and fundamental theorem that $\Phi^* = a\Phi + b\Psi$, $\Psi^* = c\Phi + d\Psi$, where a, b, c, d are real constants such that $ad - bc = -1$.[1] These constants a, b, c, d are furthermore univocally determined by the plane p. Consider next a minimal surface $\tilde{S}$ obtained from S by a rigid motion. Since a rigid motion can be obtained by combining an even number of reflections upon properly chosen planes, it follows that $\tilde{S}$ corresponds to a couple of functions $\tilde{\Phi} = \alpha\Phi + \beta\Psi$, $\tilde{\Psi} = \gamma\Phi + \delta\Psi$, where $\alpha, \beta, \gamma, \delta$ are real constants, univocally determined by the rigid motion, such that $\alpha\delta - \beta\gamma = +1$.

V.2. Suppose now that the minimal surface S, given by the formulas (5.1), is bounded by a polygon $\mathfrak{p}$, with n vertices $P_1, P_2, \ldots, P_n$. Suppose also that the correspondence between S and the half-plane $v > 0$ remains one-to-one and continuous on the boundary. Denote by $w_1, w_2, \ldots, w_n$ the images of $P_1, P_2, \ldots, P_n$. Suppose also that Φ, Ψ remain analytic on the open segments of the u-axis bounded by these images. Reflect S upon the side $P_{k-1} P_k$. The resulting surface S_k^- is then, on account of the principle of symmetry (see II.23), an analytic continuation of S, and it follows that Φ, Ψ admit of an analytic continuation through the segment $w_{k-1} w_k$ into the lower half-plane. Denote by Φ_k^-, Ψ_k^- the resulting functions. The same process, applied to the side $P_k P_{k+1}$, leads to a surface S_k^+ and to a couple of functions Φ_k^+, Ψ_k^+ in the lower half-plane. Since S_k^- and S_k^+ can be obtained from each other by a rigid motion, we have a linear relation between the couples Φ_k^-, Ψ_k^- and Φ_k^+, Ψ_k^+, which permits us to conclude that if the original couple Φ_k, Ψ_k is continued along a closed curve enclosing w_k, then there results a new couple $\Phi_k = \alpha_k\Phi + \beta_k\Psi$, $\Psi_k = \gamma_k\Phi + \delta_k\Psi$, where the real constants $\alpha_k, \beta_k, \gamma_k, \delta_k$ satisfy $\alpha_k\delta_k - \beta_k\gamma_k = +1$ and are univocally determined by the directions of the sides $P_{k-1}P_k$ and $P_k P_{k+1}$ of the polygon $\mathfrak{p}$.

V.3. Thus the couple Φ, Ψ in the formulas (5.1) undergo, by continuation along closed curves around $w_1, \ldots, w_n$, substitutions which are perfectly determined by the directions of the sides of the polygon $\mathfrak{p}$. Further conditions on Φ, Ψ can be conveniently expressed in terms of the differential equation

$$\lambda'' + p\lambda' + q\lambda = 0$$

with coefficients

$$p = -\frac{\Phi\Psi'' - \Psi\Phi''}{\Phi\Psi' - \Psi\Phi'}, \quad q = \frac{\Phi'\Psi'' - \Psi'\Phi''}{\Phi\Psi' - \Psi\Phi'},$$

which admits of the solutions $\lambda = \Phi$, $\lambda = \Psi$. It follows, from the behavior of Φ, Ψ in the vicinity of $w_1, \ldots, w_n$, that p, q are single-

[1] DARBOUX: Théorie générale des surfaces Vol. 1 Chapter 13.

valued in the whole plane and have only a finite number of poles as singularities. Hence p, q are rational functions of w. From the condition that the minimal surface S is bounded by a polygon, it follows further that p, q are both real on the w-axis[1].

V.4. These considerations suggest the following plan of attack. Determine first a differential equation $\lambda'' + p\lambda' + q\lambda = 0$, with rational coefficients $p(w)$, $q(w)$, which admits of a couple of solutions $\Phi(w)$, $\Psi(w)$ which undergo, for continuations around n points $w_1, w_2, \ldots, w_n$ given on the u-axis, given substitutions, these substitutions being perfectly determined by the directions of the sides of the given polygon $\mathfrak{p}$. The coefficients p, q must be real on the u-axis. The points $w_1, w_2, \ldots, w_n$ are, on the other hand, not univocally determined by the directions of the sides of $\mathfrak{p}$. Determine these points $w_1, w_2, \ldots, w_n$ in such a way that the minimal surface S, corresponding to Φ, Ψ by means of the formulas (5.1), be bounded by a polygon the sides of which have also the same lengths as the sides of the given polygon $\mathfrak{p}$. This is the program outlined by WEIERSTRASS and DARBOUX, and carried out by R. GARNIER.

V.5. The first step of this program requires the solution (in somewhat restricted form) of the so-called problem of RIEMANN on differential equations with prescribed group of monodromy[2]. The second step requires a delicate investigation of the behavior of the solution of the problem of RIEMANN in its dependence upon the given points $w_1, w_2, \ldots, w_n$. Thus exactly the fine points of the work of GARNIER are beyond the scope of the present report, and we have to restrict ourselves to a few remarks concerning the generality of his result.

V.6. The solution of the problem of RIEMANN secures the existence of a couple Φ, Ψ such that the formulas (5.1) determine a minimal surface S bounded by a polygon $\mathfrak{p}$ the sides of which have the prescribed directions. The next problem is to determine the points $w_1, w_2, \ldots, w_n$ in such a way that the sides of $\mathfrak{p}$ have the prescribed lengths. Suppose this problem is known to be solvable for a certain value of n, the other data of the problem being arbitrary. Take a polygon $\mathfrak{p}_{n+1}$ with $n+1$ vertices $P_1, P_2, \ldots, P_n, P_{n+1}$. Denote by $\mathfrak{p}_n$ the polygon with vertices $P_1, P_2, \ldots, P_n$. Let P_{n+1} move, on the side $P_1 P_{n+1}$, toward P_n into a position P^*_{n+1}. For the polygon $\mathfrak{p}_n$ the problem is solvable by assumption. From this GARNIER infers that the problem is possible for the polygon with vertices $P_1, P_2, \ldots, P_n, P^*_{n+1}$ if P^*_{n+1} is sufficiently close to P_n. Next he shows that the solution cannot cease to exist if P^*_{n+1}

[1] A beautiful presentation of the theory, on the basis of the older literature, is given in the classical work of DARBOUX: Théorie générale des surfaces Vol. 1 Chapter 13.

[2] See, also for literature, R. GARNIER: Solution du problème de RIEMANN. Ann. École norm. Vol. 43 (1926) pp. 177—307.

moves away from P_n on the side $P_n P_{n+1}$. It follows thus that the solution exists for $\mathfrak{p}_{n+1}$ if it is known to exist for every $\mathfrak{p}_n$. Thus it is sufficient to solve the problem for $n = 4$, in which case GARNIER shows that the existence of the solution can be verified in a direct way.

V.7. The above rough description is sufficient to explain an important fact. Suppose we apply the reasoning of GARNIER to a *knotted* polygon $\mathfrak{p}$. Then we have to deform $\mathfrak{p}$ continuously so as to obtain finally a quadrilateral, and consequently we forcibly pass through a position in which the polygon intersects itself. Well then, for polygons with multiple points problem P_3 is generally impossible. Suppose, for instance, the polygon $\mathfrak{p}$ is in the xy-plane and has the shape of an 8. Then problem P_3 reduces (see III.18) to the problem of determining a function $f(w)$ with the following properties.

1. $f(w)$ is analytic in $|w|<1$.

2. $f(w)$ is continuous in $|w|\leq 1$, and the equation $x + iy = f(w)$ carries $|w| = 1$ into the polygon $\mathfrak{p}$.

The least restrictive interpretation, consistent with the purposes of GARNIER, of condition 2 would be this: every point of $|w| = 1$ is carried into a point on $\mathfrak{p}$, and every point on $\mathfrak{p}$ is image of at least one point of $|w| = 1$. The function $f(w)$ is then not constant, and hence the image D of $|w| < 1$ is a domain (connected open set), such that $\mathfrak{p}$ is the complete boundary of D. If D_1 and D_2 are the two bounded domains bounded by the 8-shaped polygon $\mathfrak{p}$, then every point of D is in D_1 or D_2, and both D_1 and D_2 actually contain points of D. This situation contradicts the fact that D is a connected open set.

As a consequence, the method of GARNIER *gives the solution of problem P_3 only in the case of polygons which are not knotted.*

V.8. GARNIER extends his result from polygons to more general curves by a passage to the limit. Let Γ^* be a given JORDAN curve, and let $\mathfrak{p}_n$ be an inscribed polygon with n vertices. Let Φ_n, Ψ_n be the functions corresponding, in the method of GARNIER, to the polygon $\mathfrak{p}_n$. Insert a new vertex, denote by $\mathfrak{p}_{n+1}$ the resulting polygon and by Φ_{n+1}, Ψ_{n+1} the corresponding functions. Under the assumption that Γ^* consists of a finite number of arcs with bounded curvature, GARNIER is able to estimate the deviation of the couple Φ_{n+1}, Ψ_{n+1} from the couple Φ_n, Ψ_n with sufficient accuracy for the purposes of the passage to the limit. The discussion is based again upon the investigation of the analytic dependence of the solution of the problem of RIEMANN upon the data, as this dependence appears on the basis of the method developed by BIRKHOFF[1]. The fine points of the theory belong to a method which is beyond the scope of this report. As far as the geometrical

[1] Trans. Amer. Math. Soc. Vol. 10 (1909) pp. 436—470 and Vol. 12 (1911) pp. 243—284.

result is concerned, the whole discussion could be greatly simplified and generalized by applying the reasoning of V.10. It would follow in this way that *problem P_3 is solvable for every not knotted* JORDAN *curve.* It is not known at present if problem P_3 is or is not solvable for every JORDAN curve.

GARNIER also applies his method to the so-called mixed problem which calls for a minimal surface S subject to boundary conditions of the following type. The boundary Γ^* of S consists of a finite number of arcs $a_1, a_2, \ldots; b_1, b_2, \ldots$ The arcs a are given straight segments. The arcs b are not given, but they are required to be situated in given planes, and S is required to be perpendicular to these planes. The geometrical aspects of this problem are described in the classical work of DARBOUX[1]. The method of GARNIER is again based on an investigation of the solution of the problem of RIEMANN which cannot possibly be discussed in this report.

V.9. Problem P_2 (see III.5) calls for three functions $x(u, v)$, $y(u, v)$, $z(u, v)$ with the following properties.

1. $x(u, v)$, $y(u, v)$, $z(u, v)$ are harmonic in $u^2 + v^2 < 1$ and
2. satisfy there the relations $E = G$, $F = 0$, where

$$E = x_u^2 + y_u^2 + z_u^2, \quad F = x_u x_v + y_u y_v + z_u z_v, \quad G = x_v^2 + y_v^2 + z_v^2.$$

3. $x(u, v)$, $y(u, v)$, $z(u, v)$ are continuous in $u^2 + v^2 \leq 1$, and the equations $x = x(u, v)$, $y = y(u, v)$, $z = z(u, v)$ carry $u^2 + v^2 = 1$ in a one-to-one and continuous way into a given JORDAN curve Γ^*.

The methods developed for the solution of this problem work only under various additional assumptions concerning Γ^*. The following theorem is therefore of great importance.

Approximation theorem. *If a sequence Γ_n^* of* JORDAN *curves converges, in the sense of* FRÉCHET, *to a* JORDAN *curve Γ^*, and if problem P_2 is solvable for every curve of the sequence, then the problem is solvable also for the limit curve Γ^*.*

In this generality, the theorem has been proved by J. DOUGLAS[2]. T. RADÓ[3] proved the theorem under the assumption that the lengths of the curves Γ_n^* are uniformly bounded.

A JORDAN curve obviously is the limit, in the sense of FRÉCHET, of a sequence of simple closed polygons. *Hence, on account of the approximation theorem, it is sufficient to solve problem P_2 for polygons.*

V.10. The proof of the approximation theorem may be sketched as follows. Take three distinct points A, B, C on the unit circle $u^2 + v^2 = 1$, and three distinct points A^*, B^*, C^* on Γ^*. Since $\Gamma_n^* \to \Gamma^*$,

[1] Théorie générale des surfaces Vol. 1 Chapter 12.

[2] Solution of the problem of PLATEAU. Trans. Amer. Math. Soc. Vol. 33 (1931) pp. 302—306.

[3] Some remarks on the problem of PLATEAU. Proc. Nat. Acad. Sci. U. S. A. Vol. 16 (1930) pp. 242—248.

it is then possible to take three distinct points A_n^*, B_n^*, C_n^* on Γ_n^* such that $A_n^* \to A^*$, $B_n^* \to B^*$, $C_n^* \to C^*$. This being done, consider a solution $x_n(u, v)$, $y_n(u, v)$, $z_n(u, v)$ of problem P_2 for the curve Γ_n^* (the solution exists by assumption). It can be supposed that the equations

$$x = x_n(u, v), \quad y = y_n(u, v), \quad z = z_n(u, v)$$

carry the points A, B, C into the points A_n^*, B_n^*, C_n^*; we say then that we have a uniformly normalized sequence of solutions. Denote, for greater clarity, by $\xi_n(\Theta)$, $\eta_n(\Theta)$, $\zeta_n(\Theta)$ the boundary values of $x_n(u, v)$, $y_n(u, v)$, $z_n(u, v)$ on the unit circle $u = \cos\Theta$, $v = \sin\Theta$. The equations

$$x = \xi_n(\Theta), \quad y = \eta_n(\Theta), \quad z = \zeta_n(\Theta)$$

define then a one-to-one and continuous transformation T_n of $u^2 + v^2 = 1$ into Γ_n^*. All the assumptions of the selection theorem of I.22 being satisfied, there exists an everywhere convergent subsequence of these transformations. Choose a convergent subsequence and use for it the same notation as for the whole sequence. Denote by $\xi(\Theta)$, $\eta(\Theta)$, $\zeta(\Theta)$ the limits of $\xi_n(\Theta)$, $\eta_n(\Theta)$, $\zeta_n(\Theta)$. Then the equations

$$x = \xi(\Theta), \quad y = \eta(\Theta), \quad z = \zeta(\Theta)$$

define a monotonic transformation (see I.21) of $u^2 + v^2 = 1$ into a set on Γ^*. Let $x(u, v)$, $y(u, v)$, $z(u, v)$ be the harmonic functions obtained by means of the POISSON integral formula, using $\xi(\Theta)$, $\eta(\Theta)$, $\zeta(\Theta)$ as boundary functions. These harmonic functions solve then problem P_2 for the limit curve Γ^*. To see this, observe first that the sequences $\xi_n(\Theta)$, $\eta_n(\Theta)$, $\zeta_n(\Theta)$ are uniformly bounded, so that it is legitimate to make a passage to the limit under the integral sign in the formula of POISSON. Hence $x_n(u, v)$, $y_n(u, v)$, $z_n(u, v)$ and all their partial derivatives converge, in $u^2 + v^2 < 1$, to $x(u, v)$, $y(u, v)$, $z(u, v)$ and their partial derivatives respectively. From the relations

$$E_n = G_n, \quad F_n = 0 \quad \text{in} \quad u^2 + v^2 < 1$$

it follows therefore that

$$E = G, \quad F = 0 \quad \text{in} \quad u^2 + v^2 < 1.$$

It remains to investigate what happens on $u^2 + v^2 = 1$. The first step consists in proving that $\xi(\Theta)$, $\eta(\Theta)$, $\zeta(\Theta)$ are continuous. Since these functions define a monotonic transformation of $u^2 + v^2 = 1$ into Γ^*, they have definite one-sided limits $\xi^+(\Theta)$, $\eta^+(\Theta)$, $\zeta^+(\Theta)$, $\xi^-(\Theta)$, $\eta^-(\Theta)$, $\zeta^-(\Theta)$ for every Θ, and their continuity will be proved if it is shown that

$$\xi^+(\Theta) = \xi^-(\Theta), \quad \eta^+(\Theta) = \eta^-(\Theta), \quad \zeta^+(\Theta) = \zeta^-(\Theta)$$

for every Θ (see I.23). DOUGLAS does this in the following way[1]. Without loss of generality, it can be supposed that $\Theta = 0$. Take a point P on

[1] A new proof can be obtained by applying the generalized theorem of LINDELÖF, stated in III.28.

the positive u-axis, to the right of $u = 1$. Let t denote the transformation obtained by reflecting first on the circle with center at P and orthogonal to the unit circle, and reflecting second on the u-axis. t is then obviously a linear transformation carrying the unit circle into itself.

t carries therefore $x(u, v)$, $y(u, v)$, $z(u, v)$ into three functions $\bar{x}(u, v), \bar{y}(u, v), \bar{z}(u, v)$, which are harmonic in $u^2 + v^2 < 1$, and satisfy there $\bar{E} = \bar{G}$, $\bar{F} = 0$. Furthermore, if $\xi(\Theta)$, $\eta(\Theta)$, $\zeta(\Theta)$ are carried by t into $\bar{\xi}(\Theta)$, $\bar{\eta}(\Theta)$, $\bar{\zeta}(\Theta)$, then $\bar{x}(u, v)$, $\bar{y}(u, v)$, $\bar{z}(u, v)$ are given by the POISSON integral formula, with $\bar{\xi}(\Theta)$, $\bar{\eta}(\Theta)$, $\bar{\zeta}(\Theta)$ as boundary functions. Let now the point P converge to $u = 1$, and watch $\bar{\xi}(\Theta)$ for instance. Obviously, $\bar{\xi}(\Theta)$ converges to $\xi^-(0)$ on the upper half of $u^2 + v^2 = 1$, and to $\xi^+(0)$ on the lower half of $u^2 + v^2 = 1$. Consequently, $\bar{x}(u, v)$ converges in $u^2 + v^2 < 1$ to the harmonic function $x_0(u, v)$ which is obtained from the POISSON integral formula by using the boundary function $\xi_0(\Theta)$ equal to $\xi^-(0)$ for $0 < \Theta < \pi$, and equal to $\xi^+(0)$ for $\pi < \Theta < 2\pi$. The integral can then easily be computed explicitly; it follows that

$$x_0(u, v) = \Re\left(-i\frac{\xi^-(0) - \xi^+(0)}{\pi}\log\frac{1+w}{1-w} + \frac{\xi^-(0) - \xi^+(0)}{2}\right),$$

and similar expressions are obtained for the harmonic functions $y_0(u, v)$, $z_0(u, v)$ which are the limits of $\bar{y}(u, v)$, $\bar{z}(u, v)$. As $\bar{x}(u, v)$, $\bar{y}(u, v)$, $\bar{z}(u, v)$ satisfy, in $u^2 + v^2 < 1$, the relations $\bar{E} = \bar{G}$, $\bar{F} = 0$, the limit functions $x_0(u, v)$, $y_0(u, v)$, $z_0(u, v)$ must satisfy the relations $E_0 = G_0$, $F_0 = 0$. From the explicit expressions for $x_0(u, v)$, $y_0(u, v)$, $z_0(u, v)$, it follows by actual computation that

$$\left.\begin{aligned} E_0 - G_0 - 2iF_0 = &-\frac{4}{\pi^2(1-w^2)^2}[(\xi^-(0) - \xi^+(0))^2 \\ &+ (\eta^-(0) - \eta^+(0))^2 + (\zeta^-(0) - \zeta^+(0))^2]. \end{aligned}\right\}$$

Consequently, the bracket on the right-hand side vanishes. Hence, $\xi^-(0) = \xi^+(0)$, $\eta^-(0) = \eta^+(0)$, $\zeta^-(0) = \zeta^+(0)$.

The continuity of the boundary functions $\xi(\Theta)$, $\eta(\Theta)$, $\zeta(\Theta)$ being thus established, it follows that the corresponding harmonic functions remain continuous in the closed unit circle and that

$$x(u, v) = \xi(\Theta),\ y(u, v) = \eta(\Theta),\ z(u, v) = \zeta(\Theta)$$

on the perimeter of the unit circle, on account of well-known theorems concerning the POISSON integral formula.

It remains to be shown that the equations

$$x = \xi(\Theta),\ y = \eta(\Theta),\ z = \zeta(\Theta)$$

carry distinct points of $u^2 + v^2 = 1$ into distinct points of Γ^*. If this were not true, then it would follow (see I.23) that there exists an arc σ

on $u^2 + v^2 = 1$, such that $\xi(\Theta)$, $\eta(\Theta)$, $\zeta(\Theta)$ reduce all three to constants on σ. Without loss of generality, it can be supposed that $\xi(\Theta) = 0$, $\eta(\Theta) = 0$, $\zeta(\Theta) = 0$ on σ. Then[1] (principle of symmetry) the harmonic functions $x(u, v)$, $y(u, v)$, $z(u, v)$ remain analytic on σ, and consequently the relations $E = G, F = 0$ hold on σ also. Since $x(u, v) = 0$, $y(u, v) = 0$, $z(u, v) = 0$ on σ, it follows that

$$\frac{\partial x}{\partial \Theta} = -x_u \sin\Theta + x_v \cos\Theta = 0,$$

$$\frac{\partial y}{\partial \Theta} = -y_u \sin\Theta + y_v \cos\Theta = 0,$$

$$\frac{\partial z}{\partial \Theta} = -z_u \sin\Theta + z_v \cos\Theta = 0$$

on σ. Squaring and adding, we obtain $E = G = 0$ on σ, and consequently $x_u = x_v = y_u = y_v = z_u = z_v = 0$ on σ. That is to say, the functions $x_u - i x_v$, $y_u - i y_v$, $z_u - i z_v$, which are analytic functions of $w = u + iv$, vanish on a whole arc in their domain of regularity; consequently, they vanish identically. Hence, $x(u, v)$, $y(u, v)$, $z(u, v)$ and consequently the boundary functions $\xi(\Theta)$, $\eta(\Theta)$, $\zeta(\Theta)$ are constants. This contradicts however the fact that the equations $x = \xi(\Theta)$, $y = \eta(\Theta)$, $z = \zeta(\Theta)$ carry the points A, B, C on $u^2 + v^2 = 1$ into three distinct points A^*, B^*, C^* on Γ^*.

V.11. In dealing with the special case when the lengths of the curves Γ^*_n are uniformly bounded, RADÓ[2] observed that the subsequence $x_n(u, v)$, $y_n(u, v)$, $z_n(u, v)$, used in the proof of the approximation theorem, converges uniformly. According to a remark due to McSHANE (see the end of VI.10), this fact is quite general. Consider, indeed, the convergent subsequence $\xi_n(\Theta)$, $\eta_n(\Theta)$, $\zeta_n(\Theta)$, used in the proof of the approximation theorem, and denote by T_n the monotonic transformation defined by these functions. It has been proved above that the limit functions $\xi(\Theta)$, $\eta(\Theta)$, $\zeta(\Theta)$ define a continuous monotonic transformation.

Now, for sequences of monotonic functions we have the theorem that if such a sequence converges to a continuous function on a closed interval, then the convergence is uniform[3], and the proof of this theorem remains valid for sequences of monotonic transformations. Hence, $\xi_n(\Theta)$, for instance, converges uniformly to $\xi(\Theta)$, and on account of the principle of the maximum it follows that $x_n(u, v)$ converges uniformly to $x(u, v)$ in $u^2 + v^2 \leqq 1$. The approximation theorem may

[1] The following proof, taken from T. RADÓ: Some remarks on the problem of PLATEAU. Proc. Nat. Acad. Sci. U. S. A Vol. 16 (1930) pp. 242—248, is simpler than that given by DOUGLAS.

[2] Some remarks on the problem of PLATEAU. Proc. Nat. Acad. Sci. U. S. A. Vol. 16 (1930) pp. 242—248.

[3] See for instance PÓLYA-SZEGÖ: Aufgaben und Lehrsätze Vol. 1 p. 63, problem 127.

therefore be completed by the statement that the uniformly normalized sequence of the solutions of problem P_2 for the curves Γ_n^* contains a subsequence which converges *uniformly* to a solution for the limit curve Γ^*. In this form, the result may be considered as a generalization of a theorem, stated by CARATHÉODORY, on the conformal maps of sequences of plane JORDAN regions[1].

If it is known that the solution of problem P_2 for the limit curve Γ^* is unique, then it obviously follows that the whole sequence of the solutions corresponding to the curves Γ_n^* is uniformly convergent. The uniqueness theorem considered in III.11 gives examples for such situations.

V.12. We go on at present to review the methods developed for proving the existence of the solution of problem P_2. We first consider the method due to DOUGLAS[2]. Consider a monotonic transformation (see I.21)

$$T: x = \xi(\Theta), \quad y = \eta(\Theta), \quad z = \zeta(\Theta)$$

of the unit circle $u^2 + v^2 = 1$ into a set on Γ^*. The totality of these transformations constitutes the range of the argument of the functional $A(T)$, introduced by DOUGLAS, and defined by the formula

$$A(T) = \frac{1}{4\pi} \int_0^{2\pi}\int_0^{2\pi} \frac{[\xi(\Theta) - \xi(\psi)]^2 + [\eta(\Theta) - \eta(\psi)]^2 + [\zeta(\Theta) - \zeta(\psi)]^2}{4 \sin^2 \frac{\Theta - \psi}{2}} \, d\Theta \, d\psi .$$

The geometrical meaning of the integrand is as follows. Denote by l the length of the secant of the unit circle which is bounded by the points corresponding to the polar angles Θ and ψ, and denote by l^* the length of the corresponding secant of the JORDAN curve Γ^*. Then the integrand is $\left(\frac{l^*}{l}\right)^2$. This permits to show that if Γ^* is rectifiable, then there are admissible transformations T for which $A(T)$ is finite. Indeed, in this case we have a one-to-one and continuous correspondence T_0 between $u^2 + v^2 = 1$ and Γ^* such that the ratio of corresponding arcs is equal to $\frac{L}{2\pi}$, where L is the length of Γ^*. For T_0 the integrand of the functional of DOUGLAS is clearly bounded, and hence $A(T_0)$ is finite.

The method of DOUGLAS applies to JORDAN curves Γ^* such that there are admissible transformations T for which $A(T)$ is finite. The exact geometrical meaning of this assumption will be considered later; for the time being, it is sufficient to know that every rectifiable curve satisfies this condition. On account of the approximation theorem it

---

[1] See R. COURANT: Über eine Eigenschaft der Abbildungsfunktionen bei konformer Abbildung. Nachr. Ges. Wiss. Göttingen 1914 and 1922. — T. RADÓ: Sur la représentation conforme de domaines variables. Acta Litt. Sci. Szeged Vol. 1 (1923) pp. 180—186.

[2] Solution of the problem of PLATEAU. Trans. Amer. Math. Soc. Vol. 33 (1931) pp. 263—321.

would be sufficient to know that the method applies to every simple closed polygon.

V.13. Several important properties of the functional $A(T)$ become obvious when we consider another expression of $A(T)$. Using the FOURIER series of $\xi(\Theta)$, $\eta(\Theta)$, $\zeta(\Theta)$, DOUGLAS finds that $A(T)$ admits of the following equivalent definition. Let $x(u,v)$, $y(u,v)$, $z(u,v)$ denote the harmonic functions corresponding to $\xi(\Theta)$, $\eta(\Theta)$, $\zeta(\Theta)$ by means of the POISSON integral formula. Then

$$\left.\begin{aligned} A(T) &= \tfrac{1}{2}\iint\limits_{u^2+v^2<1}(E+G)\,du\,dv \\ &= \tfrac{1}{2}\Bigl[\iint\limits_{u^2+v^2<1}(x_u^2+x_v^2)\,du\,dv+\iint\limits_{u^2+v^2<1}(y_u^2+y_v^2)\,du\,dv+\iint\limits_{u^2+v^2<1}(z_u^2+z_v^2)\,du\,dv\Bigr]\end{aligned}\right\}. \quad (5.2)$$

In other words: $A(T)$ is half the sum of the DIRICHLET integrals of the harmonic functions $x(u,v)$, $y(u,v)$, $z(u,v)$.

V.14. Invariance of $A(T)$.[1] Let

$$T: x=\xi(\Theta), \quad y=\eta(\Theta), \quad z=\zeta(\Theta), \qquad (5.3)$$

be a monotonic transformation of $u^2+v^2=1$ into a set on Γ^*. Map the unit circle $u^2+v^2\leqq 1$ upon itself in a one-to-one and conformal way. The functions $\xi(\Theta)$, $\eta(\Theta)$, $\zeta(\Theta)$ are carried into new functions $\bar{\xi}(\Theta)$, $\bar{\eta}(\Theta)$, $\bar{\zeta}(\Theta)$, and

$$\bar{T}: x=\bar{\xi}(\Theta), \quad y=\bar{\eta}(\Theta), \quad z=\bar{\zeta}(\Theta)$$

is again a monotonic transformation. Two transformations T, $\bar{T}$ related in this manner are called equivalent by DOUGLAS.

If T and $\bar{T}$ are equivalent, then $A(T)=A(\bar{T})$. This is obvious from the expression (5.2) of $A(T)$, since the DIRICHLET integral is invariant under conformal mapping.

V.15. Effect of discontinuities of T on $A(T)$. Let T be given by (5.3). If $\xi(\Theta)$, for instance, is discontinuous at a certain point Θ_0, then with regard to I.23 this can happen only in the following manner: $\xi(\Theta)$ has definite one-sided limits ξ_0^+, ξ_0^- at Θ_0, and these limits are different.

In his work on conformal mapping, COURANT[2] used a simple and important reasoning which leads to the following lemma[3].

Let $\xi(\Theta)$ be integrable (in the RIEMANN sense, for instance), and let $x(u,v)$ be defined by the POISSON integral:

$$x(u,v)=\frac{1}{2\pi}\int_0^{2\pi}\xi(\Theta)\frac{1-r^2}{1+r^2-2r\cos(\Theta-\varphi)}\,d\Theta, \quad u=r\cos\varphi, \quad v=r\sin\varphi.$$

[1] The reasoning in V.14 to V.16 is somewhat simpler than that used by DOUGLAS.

[2] See HURWITZ-COURANT: Funktionentheorie, pp. 375—376.

[3] See T. RADÓ: The problem of the least area and the problem of PLATEAU. Math. Z. Vol. 32 (1930) pp. 776—779.

Suppose $\xi(\Theta)$ has definite one-sided limits ξ_0^+, ξ_0^- for some value Θ_0 of Θ, such that $\xi_0^+ \neq \xi_0^-$. Then the DIRICHLET integral

$$\iint\limits_{u^2+v^2<1} (x_u^2 + x_v^2)\, du\, dv$$

cannot be finite.

Applying this to the function $\xi(\Theta)$ figuring in the definition of the transformation T, we obtain from the expression (5.2) of $A(T)$ that if $\xi(\Theta)$ has a discontinuity, then $A(T)$ cannot be finite. The same remark applies of course to $\eta(\Theta)$, $\zeta(\Theta)$. Hence:

If $A(T)$ is finite, then the transformation T necessarily is continuous.

V.16. *Lower semi-continuity of $A(T)$.* Consider a sequence

$$T_n: x = \xi_n(\Theta), \quad y = \eta_n(\Theta), \quad z = \zeta_n(\Theta)$$

of monotonic transformations. Suppose T_n converges to a monotonic transformation

$$T: x = \xi(\Theta),\ y = \eta(\Theta),\ z = \zeta(\Theta).$$

Then

$$A(T) \leqq \underline{\lim}\, A(T_n). \tag{5.4}$$

Denote by $x_n(u, v)$, $y_n(u, v)$, $z_n(u, v)$ the harmonic functions corresponding to $\xi_n(\Theta)$, $\eta_n(\Theta)$, $\zeta_n(\Theta)$ by means of the POISSON integral formula. Let $x(u, v)$, $y(u, v)$, $z(u, v)$ have the same meaning with respect to $\xi(\Theta)$, $\eta(\Theta)$, $\zeta(\Theta)$. Then it follows from the POISSON integral formula that $x_n(u, v)$, $y_n(u, v)$, $z_n(u, v)$ and their partial derivatives converge to $x(u, v)$, $y(u, v)$, $z(u, v)$ and their partial derivatives respectively in $u^2 + v^2 < 1$, and that the convergence is uniform in every concentric circle $u^2 + v^2 \leqq r^2 < 1$. Hence, for $0 < r < 1$:

$$\tfrac{1}{2}\iint\limits_{u^2+v^2<r^2} (E + G)\, du\, dv = \lim \tfrac{1}{2}\iint\limits_{u^2+v^2<r^2} (E_n + G_n)\, du\, dv \leqq \underline{\lim}\, \tfrac{1}{2}\iint\limits_{u^2+v^2<1} (E_n + G_n)\, du\, dv.$$

The passage to the limit $r \to 1$ yields the inequality (5.4).

V.17. The existence proof of DOUGLAS proceeds then in the following steps. Let m denote the (by assumption finite) greatest lower bound of the functional $A(T)$. There exists a sequence $\overline{T}_n$ such that $A(\overline{T}_n) \to m$. Of course, only such transformations $\overline{T}_n$ will be used for which $A(\overline{T}_n)$ is finite. Hence, $\overline{T}_n$ is continuous (see V.15). Choose now three distinct points A, B, C on $u^2 + v^2 = 1$, and three distinct fixed points A^*, B^*, C^* on the given JORDAN curve Γ^*. Since $\overline{T}_n$ is continuous, A^*, B^*, C^* will be the images, under $\overline{T}_n$, of three distinct points A_n, B_n, C_n. Using a linear transformation of the unit circle into itself which carries A_n, B_n, C_n into A, B, C, we obtain a monotonic transformation T_n which carries A, B, C into A^*, B^*, C^* and which is equivalent to $\overline{T}_n$. Hence $A(T_n) = A(\overline{T}_n)$ (see V.14). In this way we obtain a sequence T_n of monotonic transformations such that $A(T_n) \to m$, and such that every T_n carries the three prescribed points A, B, C of $u^2 + v^2 = 1$

into the three prescribed points A^*, B^*, C^* of Γ^*. The sequence T_n satisfies the assumptions of the selection theorem in I.22, and therefore contains an everywhere convergent subsequence, which we denote by T_n again for simplicity. There results a limit transformation

$$T_0 : x = \xi_0(\Theta), \quad y = \eta_0(\Theta), \quad z = \zeta_0(\Theta), \tag{5.5}$$

which is again a monotonic transformation of $u^2 + v^2 = 1$ into a set on Γ^*. Obviously, T_0 carries the points A, B, C into the points A^*, B^*, C^* respectively. By the definition of m we have $A(T_0) \geqq m$, while on the other hand, on account of the lower semi-continuity of the functional of DOUGLAS, $A(T_0) \leqq \underline{\lim}\, A(T_n) = m$. Hence, $A(T_0) = m$. Since m is finite, T_0 is continuous (see V.15). Summing up:

There exists a transformation T_0 which 1. minimizes the functional $A(T)$, which 2. is continuous, and which 3. carries the three prescribed points A, B, C on $u^2 + v^2 = 1$ into three prescribed points A^*, B^*, C^* on Γ^*.

V.18. Consider this minimizing T_0, given by (5.5). Denote by $x_0(u, v)$, $y_0(u, v)$, $z_0(u, v)$ the harmonic functions which reduce to $\xi_0(\Theta)$, $\eta_0(\Theta)$, $\zeta_0(\Theta)$ on $u^2 + v^2 = 1$. x_0, y_0, z_0 are the real parts of functions Φ_1, Φ_2, Φ_3, which are analytic functions of $w = u + iv$ in $u^2 + v^2 < 1$. Then we have (cf. II.18)

$$E_0 - G_0 - 2iF_0 = \Phi_1'^2 + \Phi_2'^2 + \Phi_3'^2. \tag{5.6}$$

The next step is to show that the minimizing property of T_0 has the consequence that

$$\Phi_1'^2 + \Phi_2'^2 + \Phi_3'^2 = 0.$$

This comes out as follows[1]. By a series of computations, the following formulas are established.

$$\Phi_1'^2 + \Phi_2'^2 + \Phi_3'^2 = \frac{1}{2\pi^2} \int_0^{2\pi} \int_0^{2\pi} \Psi(\varphi, \psi) \frac{dz\, d\zeta}{(z - w)^2 (\zeta - w)^2}, \tag{5.7}$$

$$A(T_0) = \frac{1}{4\pi} \int_0^{2\pi} \int_0^{2\pi} \Psi(\varphi, \psi) \frac{dz\, d\zeta}{(z - \zeta)^2},$$

where $z = e^{i\varphi}$, $\zeta = e^{i\psi}$, and

$$\Psi(\varphi, \psi) = [\xi_0(\varphi) - \xi_0(\psi)]^2 + [\eta_0(\varphi) - \eta_0(\psi)]^2 + [\zeta_0(\varphi) - \zeta_0(\psi)]^2.$$

Let then λ be a real parameter and consider the following (for small values of λ obviously one-to-one and continuous) transformations of the unit circle $u = \cos\Theta$, $v = \sin\Theta$ into itself:

$$z' = z \exp \lambda \left[\frac{1}{z(z - w)} - \frac{1}{\bar{z}(\bar{z} - \bar{w})} \right],$$

and

$$z' = z \exp \lambda \left[\frac{1}{z(z - w)} + \frac{1}{\bar{z}(\bar{z} - \bar{w})} \right],$$

[1] We follow the simplified presentation given by DOUGLAS in a later paper: The problem of PLATEAU for two contours. J. Math. Physics, Massachusetts Inst. Technol. Vol. 10 (1931) pp. 315–359.

where the bar denotes the conjugate complex number, and w is a fixed interior point of the unit circle. Using one of these transformations, we transform $\xi_0(\Theta)$, $\eta_0(\Theta)$, $\zeta_0(\Theta)$ into three new functions $\xi(\Theta)$, $\eta(\Theta)$, $\zeta(\Theta)$; and

$$T: x = \xi(\Theta),\ y = \eta(\Theta),\ z = \zeta(\Theta)$$

is a monotonic transformation which depends upon λ. Consequently $A(T)$ is a function $J(\lambda)$ of λ which has a minimum for $\lambda = 0$. Hence $J'(0) = 0$.

Computation shows that

$$J'(0) = -\frac{1}{4\pi}\int_0^{2\pi}\int_0^{2\pi} \Psi(\varphi, \psi)\,\frac{dz\,d\zeta}{(z-w)^2(\zeta-w)^2}.$$

Comparing with (5.7), we see that $J'(0) = 0$ implies that $\Phi_1'^2 + \Phi_2'^2 + \Phi_3'^2 = 0$. Consequently, on account of (5.6), $E_0 = G_0$, $F_0 = 0$.

Thus, the harmonic functions $x_0(u, v)$, $y_0(u, v)$, $z_0(u, v)$ satisfy the relations $E_0 = G_0$, $F_0 = 0$ for $u^2 + v^2 < 1$. They are continuous in $u^2 + v^2 \leqq 1$. On $u^2 + v^2 = 1$ they reduce to the functions $\xi(\Theta)$, $\eta(\Theta)$, $\zeta(\Theta)$, and consequently the equations

$$x = x_0(u, v),\quad y = y_0(u, v),\quad z = z_0(u, v)$$

carry $u^2 + v^2 = 1$ in a continuous and monotonic way into Γ^*, the points A, B, C of $u^2 + v^2 = 1$ being taken into the prescribed points A^*, B^*, C^* of Γ^*. It remains to be shown that distinct points of $u^2 + v^2 = 1$ are taken into distinct points of Γ^*. If this were not the case, there would exist an arc of $u^2 + v^2 = 1$ on which $x_0(u, v)$, $y_0(u, v)$, $z_0(u, v)$ would all three reduce to constants. Since $E_0 = G_0$, $F_0 = 0$, it would then follow (cf. the end of V.10) that $x_0(u, v)$, $y_0(u, v)$, $z_0(u, v)$ reduce to constants identically. This contradicts however the fact that the points A, B, C are carried into three distinct points A^*, B^*, C^*.

Thus the functions $x_0(u, v)$, $y_0(u, v)$, $z_0(u, v)$ solve problem P_2 for the given JORDAN curve Γ^*. The generality of Γ^* is restricted by the assumption that the greatest lower bound of the functional $A(T)$ is finite. This assumption is however certainly satisfied for every rectifiable JORDAN curve (see V.12), and hence, in particular, for every simple closed polygon. On account of the approximation theorem, this is sufficient to establish the existence of the solution of problem P_2 for every JORDAN curve.

V.19. Problem P_2, as stated above, is considered by DOUGLAS as the three-dimensional special case of the following *n-dimensional problem*. Determine n functions $x_1(u, v), \ldots, x_n(u, v)$ with the following properties.

1. $x_1(u, v), \ldots, x_n(u, v)$ are harmonic for $u^2 + v^2 < 1$ and
2. satisfy there the relations $E = G$, $F = 0$, where

$$E = x_{1u}^2 + \cdots + x_{nu}^2,\quad F = x_{1u}x_{1v} + \cdots + x_{nu}x_{nv},\quad G = x_{1v}^2 + \cdots + x_{nv}^2.$$

3. $x_1(u, v), \ldots, x_n(u, v)$ are continuous in $u^2 + v^2 \leqq 1$ and the equations $x_1 = x_1(u, v), \ldots, x_n = x_n(u, v)$ carry $u^2 + v^2 = 1$ in a one-to-one and continuous way into a JORDAN curve Γ^* given in the n-dimensional space.

DOUGLAS carries out his method for a general n and emphasizes the fact that the value of n does not make any difference. For this reason we restricted ourselves, in the preceding review of his method, to the familiar three-dimensional case. Besides the case $n = 3$, the case $n = 2$ is important. Let us first recall the classical *theorem of* DARBOUX[1]. Let R be a closed region in the $w = u + iv$ plane, bounded by a JORDAN curve. Let there be given a function $f(w)$, which is continuous in R and analytic in the interior of R. Suppose that $f(w)$ takes on distinct values at distinct boundary points of R, that is to say suppose that the equation $\zeta = f(w)$ carries the boundary curve of R in a one-to-one and continuous way into a JORDAN curve in the ζ-plane. Denote by R^* the region bounded by this image curve. *The equation $\zeta = f(w)$ carries then R in a one-to-one and continuous way into R^*, and this transformation is conformal in the interior of the regions.*

Consider now the case $n = 2$ of problem P_2. We have then two harmonic functions $x_1(u, v)$, $x_2(u, v)$, and condition 2 of the problem implies that $x_1(u, v)$, $x_2(u, v)$ are conjugate harmonic functions, that is to say real and imaginary part respectively of an analytic function $f(w)$ of $w = u + iv$. On account of the theorem of DARBOUX, it follows that in the case $n = 2$, problem P_2 reduces to the problem of finding a function $\zeta = f(w)$, which maps the unit circle $|w| \leqq 1$ upon a given JORDAN region in a one-to-one, continuous and, in the interior of the regions, conformal way. DOUGLAS points out that his method is entirely independent of the theory of conformal mapping and that consequently his method yields a new solution of this important problem. He also points out that while the modern methods of dealing with the problem establish first the existence of the mapping function for the interior and obtain information as to its behavior on the boundary afterwards, his method leads directly to the correspondence of the boundaries. The behavior of the mapping function in the interior follows then immediately from the theorem of DARBOUX.

Thus the fact that the method of DOUGLAS is independent of the theory of conformal mapping appears as one of its essential features.

V.20. The method of T. RADO[2], which will be reviewed presently, depends essentially on the theory of conformal mapping. The idea of the method is to construct first an *approximate solution of the problem,* and to obtain the exact solution by a *passage to the limit.* The con-

[1] See for instance OSGOOD: Lehrbuch der Funktionentheorie Vol. 1, 5. edition pp. 397—399.

[2] On PLATEAU's problem. Ann. of Math. Vol. 31 (1930) pp. 457—469.

struction of the approximate solution being the essential part of the method, it may be interesting to point out the simple facts on which the construction is based. Let $\mathfrak{a}(\Gamma^*)$ denote the greatest lower bound of the areas of all the surfaces (of the type of the circle) bounded by a given JORDAN curve Γ^*. Given then a $\sigma > 0$, there exists a surface

$$\overline{S}: x = \bar{x}(u, v),\ y = \bar{y}(u, v),\ z = \bar{z}(u, v),\ u^2 + v^2 \leqq 1 \qquad (5.8)$$

bounded by Γ^* and such that

$$\mathfrak{a}(\Gamma^*) \leqq \mathfrak{A}(\overline{S}) \leqq \mathfrak{a}(\Gamma^*) + \sigma,$$

where $\mathfrak{A}(\overline{S})$ denotes the area of $\overline{S}$. Make now the (generally unjustified) assumption that $\overline{S}$ admits of a conformal map. Suppose that (5.8) is such a map; then $\overline{E} = \overline{G}$, $\overline{F} = 0$, and the formula

$$\mathfrak{A}(\overline{S}) = \iint (\overline{E}\overline{G} - \overline{F}^2)^{\frac{1}{2}}$$

reduces to

$$\mathfrak{A}(\overline{S}) = \iint \overline{E} = \iint \overline{G} = \tfrac{1}{2} \iint (\overline{E} + \overline{G}),$$

where the integrals are taken in $u^2 + v^2 < 1$. Denote then by $x(u, v)$, $y(u, v)$, $z(u, v)$ the harmonic functions which coincide, for $u^2 + v^2 = 1$, with $\bar{x}(u, v)$, $\bar{y}(u, v)$, $\bar{z}(u, v)$ respectively. Then

$$S: x = x(u, v),\ y = y(u, v),\ z = z(u, v),\ u^2 + v^2 \leqq 1$$

is again bounded by Γ^* and consequently $\mathfrak{a}(\Gamma^*) \leqq \mathfrak{A}(S)$. Since a harmonic function, with prescribed boundary values, minimizes the DIRICHLET integral, we have

$$\left.\begin{aligned} \iint (x_u^2 + x_v^2) &\leqq \iint (\bar{x}_u^2 + \bar{x}_v^2), \\ \iint (y_u^2 + y_v^2) &\leqq \iint (\bar{y}_u^2 + \bar{y}_v^2), \\ \iint (z_u^2 + z_v^2) &\leqq \iint (\bar{z}_u^2 + \bar{z}_v^2). \end{aligned}\right\} \qquad (5.9)$$

Addition gives

$$\iint (E + G) \leqq \iint (\overline{E} + \overline{G}).$$

Combining all the preceding information, we can write

$$\left.\begin{aligned} \mathfrak{a}(\Gamma^*) \leqq \mathfrak{A}(S) = \iint (EG - F^2)^{\frac{1}{2}} \leqq \iint E^{\frac{1}{2}} G^{\frac{1}{2}} \leqq \tfrac{1}{2} \iint (E + G) \\ \leqq \tfrac{1}{2} \iint (\overline{E} + \overline{G}) = \mathfrak{A}(\overline{S}) \leqq \mathfrak{a}(\Gamma^*) + \sigma. \end{aligned}\right\} \qquad (5.10)$$

Hence any two of the integrals appearing in this sequence of inequalities differ from each other by not more than σ, and this leads immediately to the inequalities

$$\iint (E^{\frac{1}{2}} - G^{\frac{1}{2}})^2 \leqq 2\sigma,$$

$$\iint |F| \leqq [2\sigma(\sigma + \mathfrak{a}(\Gamma^*))]^{\frac{1}{2}}.$$

Thus the functions $x(u, v)$, $y(u, v)$, $z(u, v)$ satisfy the conditions of problem P_2, except for the conditions $E = G$, $F = 0$. These conditions are satisfied *approximately* in the sense that the integrals

$$\iint (E^{\frac{1}{2}} - G^{\frac{1}{2}})^2 \quad \text{and} \quad \iint |F|$$

are as small as we please, since σ was arbitrary.

The case $\sigma = 0$ explains the idea and the origin of this construction[1]. If $\sigma = 0$, then we have the sign of equality all over in (5.10), and consequently also in (5.9). This implies however that $x(u, v) \equiv \bar{x}(u, v)$, $y(u, v) \equiv \bar{y}(u, v)$, $z(u, v) \equiv \bar{z}(u, v)$. That is to say, $\bar{x}(u, v)$, $\bar{y}(u, v)$, $\bar{z}(u, v)$ are harmonic functions. *Hence, if we have a surface $\bar{S}$ which admits of a conformal map and whose area is a minimum, then $\bar{S}$ solves the problem of* PLATEAU. *The above construction shows then that if we have a surface S which admits of a conformal map and whose area is approximately a minimum, then we can derive from it an approximate solution of the problem of* PLATEAU.

This construction of the approximate solution is based on the assumption that the surface S admits of a conformal map. In general, however, a surface does *not* admit of a conformal map. On the other hand, there is a great latitude in the choice of S. Further discussion will show that $\bar{S}$ can be replaced, as far as the successful application of the construction is concerned, by a nearby *polyhedron*. Thus it is sufficient to know that polyhedrons admit of conformal maps, and this already has been established by H. A. SCHWARZ. We now are going to describe briefly the actual application of the preceding considerations to the solution of problem P_2.

V.21. Suppose the given JORDAN curve is a simple closed polygon and denote this polygon by $\mathfrak{p}^*$. Consider all the polyhedrons bounded by $\mathfrak{p}^*$ (see I.9) and denote by $\mu(\mathfrak{p}^*)$ the greatest lower bound of their areas. $\mu(\mathfrak{p}^*)$ obviously is finite.

Consider also all the continuous surfaces bounded by $\mathfrak{p}^*$ and denote by $\mathfrak{a}(\mathfrak{p}^*)$ the greatest lower bound of their areas. Obviously, $\mathfrak{a}(\mathfrak{p}^*) \leqq \mu(\mathfrak{p}^*)$. In his Thesis[2], LEBESGUE proved a theorem from which it follows that $\mathfrak{a}(\mathfrak{p}^*) = \mu(\mathfrak{p}^*)$. This means that for the area $\mathfrak{A}(S)$ of every continuous surface S, bounded by $\mathfrak{p}^*$, we have $\mathfrak{A}(S) \geqq \mu(\mathfrak{p}^*)$. The result of LEBESGUE, referred to above, escaped the attention of T. RADÓ. He makes use of the fact, easily established, that the inequality $\mathfrak{A}(S) \geqq \mu(\mathfrak{p}^*)$ holds for harmonic surfaces bounded by $\mathfrak{p}^*$, that is to say for surfaces which admit of a representation $\mathfrak{x} = \mathfrak{x}(u, v)$, $u^2 + v^2 \leqq 1$, where the components $x(u, v)$, $y(u, v)$, $z(u, v)$ of $\mathfrak{x}(u, v)$ are harmonic functions.

[1] See II.10.

[2] Intégrale, longueur, aire. Ann. Mat. pura appl. Vol. 7 (1902) pp. 231—359.

V.22. The first step is then to solve the following approximate problem. Give three distinct points A, B, C on $u^2+v^2=1$, three distinct points A^*, B^*, C^* on $\mathfrak{p}^*$, and also give an $\varepsilon > 0$. Determine three functions $x(u,v)$, $y(u,v)$, $z(u,v)$ with the following properties.

1. $x(u,v)$, $y(u,v)$, $z(u,v)$ are harmonic for $u^2+v^2<1$, and
2. satisfy the relations

$$\iint (E^{\frac{1}{2}} - G^{\frac{1}{2}})^2 \leqq \varepsilon, \quad \iint |F| \leqq \varepsilon,$$

where the integrals are extended over $u^2+v^2<1$.

3. $x(u,v)$, $y(u,v)$, $z(u,v)$ are continuous in $u^2+v^2 \leqq 1$, and the equations $x = x(u,v)$, $y = y(u,v)$, $z = z(u,v)$ carry $u^2+v^2=1$ in a one-to-one and continuous way into $\mathfrak{p}^*$, the points A, B, C being carried into the points A^*, B^*, C^*.

This problem is solved as follows. Let $\sigma > 0$ be a constant to be determined later on. By the definition of $\mu(\mathfrak{p}^*)$ (see V.21) there is a polyhedron $\overline{\mathfrak{P}}$, bounded by $\mathfrak{p}^*$, such that $\mathfrak{A}(\overline{\mathfrak{P}}) \leqq \mu(\mathfrak{p}^*) + \sigma$. Let

$$\overline{\mathfrak{P}}: x = \bar{x}(u,v), \ y = \bar{y}(u,v), \ z = \bar{z}(u,v), \ u^2+v^2 \leqq 1$$

be an isothermic representation of $\overline{\mathfrak{P}}$, such that A, B, C are carried into A^*, B^*, C^*. We have then, on account of $\overline{E} = \overline{G}, \overline{F} = 0$,

$$\mathfrak{A}(\overline{\mathfrak{P}}) = \tfrac{1}{2}\iint (\overline{E} + \overline{G}).$$

Let $x(u,v)$, $y(u,v)$, $z(u,v)$ be the harmonic functions which coincide with $\bar{x}(u,v)$, $\bar{y}(u,v)$, $\bar{z}(u,v)$ respectively on $u^2+v^2=1$. *Then these harmonic functions solve the approximate problem.*

Indeed, conditions 1 and 3 obviously are satisfied. To verify that condition 2 is satisfied, use again the fact that a harmonic function with given boundary values minimizes the DIRICHLET integral. This gives (cf. V.20)

$$\iint (E+G) \leqq \iint (\overline{E} + \overline{G}).$$

The surface

$$S: x = x(u,v), \ y = y(u,v), \ z = z(u,v), \ u^2+v^2 \leqq 1$$

is a harmonic surface bounded by $\mathfrak{p}^*$. Hence (see V.21)

$$\mu(\mathfrak{p}^*) \leqq \mathfrak{A}(S) = \iint (EG - F^2)^{\frac{1}{2}}.$$

Combining all the preceding information, we can write

$$\left.\begin{aligned} \mu(\mathfrak{p}^*) \leqq \mathfrak{A}(S) &= \iint (EG - F^2)^{\frac{1}{2}} \leqq \iint E^{\frac{1}{2}} G^{\frac{1}{2}} \leqq \tfrac{1}{2}\iint (E+G) \\ &\leqq \tfrac{1}{2}\iint (\overline{E} + \overline{G}) = \mathfrak{A}(\overline{\mathfrak{P}}) \leqq \mu(\mathfrak{p}^*) + \sigma. \end{aligned}\right\}$$

Hence (cf. V.20)

$$\iint (E^{\frac{1}{2}} - G^{\frac{1}{2}})^2 \leqq 2\sigma,$$

$$\iint |F| \leqq [2\sigma(\sigma + \mu(\mathfrak{p}^*))]^{\frac{1}{2}}.$$

By proper choice of σ these bounds can be made less than the prescribed ε.

V.23. The solution of the exact problem P_2 is obtained now by a passage to the limit. Keep the points A, B, C and A^*, B^*, C^* fixed and solve the approximate problem for $\varepsilon_1, \varepsilon_2, \ldots, \varepsilon_n, \ldots$, where $\varepsilon_n \to 0$. There results a sequence of approximate solutions $x_n(u, v)$, $y_n(u, v)$, $z_n(u, v)$ and the problem is to show that a properly chosen subsequence converges to a solution of the exact problem.

Suppose we have obtained, in any manner whatsoever, a subsequence $x_{n_k}(u, v)$, $y_{n_k}(u, v)$, $z_{n_k}(u, v)$ converging in $u^2 + v^2 < 1$ to the (necessarily harmonic) functions $x(u, v)$, $y(u, v)$, $z(u, v)$. The convergence extends then to all partial derivatives, and is uniform in every concentric circle $u^2 + v^2 \leqq r^2 < 1$. Hence, r being fixed,

$$\iint\limits_{(r)} (E_n^{\frac{1}{2}} - G_n^{\frac{1}{2}})^2 \to \iint\limits_{(r)} (E^{\frac{1}{2}} - G^{\frac{1}{2}})^2, \quad \iint\limits_{(r)} |F_n| \to \iint\limits_{(r)} |F|,$$

where (r) indicates that the integrals are taken over $u^2 + v^2 \leqq r^2$. Since

$$\iint\limits_{(r)} (E_n^{\frac{1}{2}} - G_n^{\frac{1}{2}})^2 \leqq \iint\limits_{u^2+v^2<1} (E_n^{\frac{1}{2}} - G_n^{\frac{1}{2}})^2 \leqq \varepsilon_n \to 0,$$

$$\iint\limits_{(r)} |F_n| \leqq \iint\limits_{u^2+v^2<1} |F_n| \leqq \varepsilon_n \to 0,$$

it follows that

$$\iint\limits_{(r)} (E^{\frac{1}{2}} - G^{\frac{1}{2}})^2 = 0, \quad \iint\limits_{(r)} |F| = 0.$$

Since E, F, G are continuous, it follows that $E = G$, $F = 0$ for $u^2 + v^2 < r^2$. Since this is true for every $r < 1$, it follows that $E = G$, $F = 0$ in the whole interior of the unit circle.

Using this simple remark, T. RADÓ verifies that the method he used to prove a special case of the approximation theorem (see V.9), namely the special case when the lengths of the approximating curves are uniformly bounded, works without further modification under the present conditions. The reason for this is the fact that the approximate relations

$$\iint (E_n^{\frac{1}{2}} - G_n^{\frac{1}{2}})^2 \leqq \varepsilon_n, \quad \iint |F_n| \leqq \varepsilon_n$$

are, for the purposes of the passage to the limit, just as efficient as the exact relations $E = G$, $F = 0$ would be.

In this way follows the existence of the solution of problem P_2 for every simple closed polygon. Using the special case of the approximation theorem he has previously proved, T. RADÓ extends then the existence theorem to rectifiable JORDAN curves.

V.24. The reviewer takes the liberty to call the attention of the reader to a very simple and efficient method resulting from a combination of the proof of the approximation theorem of J. DOUGLAS with the notion and the construction, given above, of the approximate solution. Let the JORDAN curves Γ_n^* converge, in the sense of FRÉCHET, to the JORDAN curve Γ^*. Take three distinct points A, B, C on $u^2 + v^2 = 1$, three distinct points A^*, B^*, C^* on Γ^*, and three distinct points A_n^*, B_n^*, C_n^* on Γ_n^* in such a way that $A_n^* \to A^*$, $B_n^* \to B^*$, $C_n^* \to C^*$ (this is possible since $\Gamma_n^* \to \Gamma^*$). Give also a sequence of positive constants $\varepsilon_n \to 0$. Suppose that it is possible to solve, for every curve of the sequence Γ_n^*, the following approximate problem. Determine three functions $x_n(u, v)$, $y_n(u, v)$, $z_n(u, v)$ with the following properties.

1. $x_n(u, v)$, $y_n(u, v)$, $z_n(u, v)$ are harmonic in $u^2 + v^2 < 1$, and
2. satisfy the relations

$$\iint (E_n^{\frac{1}{2}} - G_n^{\frac{1}{2}})^2 \leqq \varepsilon_n, \quad \iint |F_n| \leqq \varepsilon_n.$$

3. $x_n(u, v)$, $y_n(u, v)$, $z_n(u, v)$ are continuous in $u^2 + v^2 \leqq 1$, and the equations $x = x_n(u, v)$, $y = y_n(u, v)$, $z = z_n(u, v)$ carry $u^2 + v^2 = 1$ in a one-to-one and continuous way into Γ_n^*, the points A, B, C being carried into A_n^*, B_n^*, C_n^*.

Apply then to this situation the proof of the approximation theorem of J. DOUGLAS, supplemented by the remarks of V.23 concerning the handling of the approximate relations

$$\iint (E_n^{\frac{1}{2}} - G_n^{\frac{1}{2}})^2 \leqq \varepsilon_n, \quad \iint |F_n| \leqq \varepsilon_n.$$

There follows, without any further change in the proof, that there exists a subsequence converging uniformly, in $u^2 + v^2 \leqq 1$, to an exact solution of problem P_2 for the limit curve Γ^*.

Start then with a given JORDAN curve Γ^* and approximate it, in the sense of FRÉCHET, by a sequence of simple closed polygons $\mathfrak{p}_n^*$. Use the method of V.22 to solve the approximate problem for $\mathfrak{p}_n^*$. The generalization of the approximation theorem, described above, yields then the solution of the exact problem P_2 for the given JORDAN curve Γ^*.

V.25. One of the important steps in the method of J. DOUGLAS is the discussion of the first variation of his functional. T. RADÓ observed that this discussion can be based on a simple lemma concerning the conformal maps of surfaces[1].

Let $\mathfrak{A}(S)$ denote the area of a (sufficiently regular) surface

$$S: x = x(u, v), \quad y = y(u, v), \quad z = z(u, v), \quad u^2 + v^2 \leqq 1. \tag{5.11}$$

Then
$$\mathfrak{A}(S) = \iint (EG - F^2)^{\frac{1}{2}} \leqq \iint E^{\frac{1}{2}} G^{\frac{1}{2}} \leqq \tfrac{1}{2} \iint (E + G). \tag{5.12}$$

Hence
$$\mathfrak{A}(S) \leqq \tfrac{1}{2} \iint (E + G)$$

for every representation of S, while for isothermic representations obviously

$$\mathfrak{A}(S) = \tfrac{1}{2} \iint (E + G),$$

on account of $E = G$, $F = 0$. While hasty inferences from these remarks are invalidated by the fact that a surface, in general, does not admit

[1] T. RADÓ: On the functional of Mr. DOUGLAS. Ann. of Math. Vol. 32 (1931) pp. 785—803.

of an isothermic representation (see I.18), still it is possible to draw a conclusion sufficiently strong for the purposes of the problem of PLATEAU.

Suppose a surface S admits of a representation (5.11) where $x(u, v)$, $y(u, v)$, $z(u, v)$ are continuous for $u^2 + v^2 \leqq 1$ and have continuou- partial derivatives of the first and second order for $u^2 + v^2 < 1$. Ins troduce new parameters $\bar{u}, \bar{v}$ by equations $u = u(\bar{u}, \bar{v})$, $v = v(\bar{u}, \bar{v})$. The parameters $\bar{u}, \bar{v}$ will be called *admissible parameters* if the following conditions are satisfied.

1. The equations $u = u(\bar{u}, \bar{v})$, $v = v(\bar{u}, \bar{v})$ carry $\bar{u}^2 + \bar{v}^2 \leqq 1$ in a one-to-one and continuous way into $u^2 + v^2 \leqq 1$.

2. The functions $u(\bar{u}, \bar{v})$, $v(\bar{u}, \bar{v})$ have continuous partial derivatives of the first and second order for $\bar{u}^2 + \bar{v}^2 < 1$.

3. The JACOBIAN $\partial(u, v)/\partial(\bar{u}, \bar{v})$ is different from zero for $\bar{u}^2 + \bar{v}^2 < 1$.

Changing to admissible parameters $\bar{u}, \bar{v}$, we obtain the equations of S in the form

$$S: x = \bar{x}(\bar{u}, \bar{v}), \quad y = \bar{y}(\bar{u}, \bar{v}), \quad z = \bar{z}(\bar{u}, \bar{v}), \quad \bar{u}^2 + \bar{v}^2 \leqq 1 .$$

The representations of S, obtained in this way, will be called *admissible representations*.

For every admissible representation $S: \mathfrak{x} = \mathfrak{x}(u, v)$, $u^2 + v^2 \leqq 1$, consider the integral

$$J(R) = \frac{1}{2} \iint_{u^2+v^2<1} (E + G)\, du\, dv ,$$

where R denotes the admissible representation which has been used to compute the integral.

This functional $J(R)$ is defined by the same integral as the functional $A(T)$ of J. DOUGLAS (see V.13). It should be remembered, however, that the argument of $A(T)$ is a variable representation T of a given JORDAN curve, while the argument R of $J(R)$ is a variable representation of a given surface.

Suppose now that for the given surface S the greatest lower bound of $J(R)$ is finite. Then we have the following

Lemma. *If the variation problem $J(R)$ = minimum has a solution R, then R is an isothermic representation of S. Conversely, if S admits of an isothermic representation R, then R is a solution of the variation problem $J(R)$ = minimum.*

The second half of the lemma is trivial. Indeed, we have for every admissible representation R the relations (5.12). Hence $\mathfrak{A}(S) \leqq J(R)$ for every admissible representation, while for isothermic representations we have obviously $\mathfrak{A}(S) = J(R)$.

The first half of the lemma would also be trivial if we would know a priori that S admits of isothermic representations. Indeed, if R_0 is an isothermic representation, then $\mathfrak{A}(S) = J(R_0)$, while $\mathfrak{A}(S) \leqq J(R)$

for all admissible representations. Hence $\min J(R) = \mathfrak{A}(S)$. If then a representation R minimizes $J(R)$, we must have $\mathfrak{A}(S) = J(R)$. It follows that in the relations (5.12) we must have the sign of equality all over, which obviously implies that $E = G$, $F = 0$.

The point therefore is that the lemma is true without any previous assumption concerning the existence of isothermic representations. The proof of the lemma runs as follows. Let

$$S: x = x(u, v), \quad y = y(u, v), \quad z = z(u, v), \quad u^2 + v^2 \leqq 1$$

be the minimizing representation. Let $\alpha = \alpha(u, v)$, $\beta = \beta(u, v)$ be admissible parameters, and let

$$S: x = \bar{x}(\alpha, \beta), \; y = \bar{y}(\alpha, \beta), \; z = \bar{z}(\alpha, \beta), \; \alpha^2 + \beta^2 \leqq 1$$

be the resulting admissible representation $\bar{R}$. Then $J(\bar{R})$ is an integral over $\alpha^2 + \beta^2 < 1$. Introducing in this integral the old parameters u, v as variables of integration, we obtain the formula

$$J(\bar{R}) = \frac{1}{2} \iint\limits_{u^2+v^2<1} [E(\alpha_v^2 + \beta_v^2) - 2F(\alpha_u \alpha_v + \beta_u \beta_v) + G(\alpha_u^2 + \beta_u^2)] \frac{\partial(u, v)}{\partial(\alpha, \beta)} \, du \, dv. \quad (5.13)$$

Let ε be a small parameter, and let $\psi(u, v)$ be a function which is continuous and has continuous partial derivatives of the first and second order in and on the unit circle $u^2 + v^2 = 1$. Then the equations

$$\alpha = u \cos \varepsilon\psi - v \sin \varepsilon\psi,$$
$$\beta = u \sin \varepsilon\psi + v \cos \varepsilon\psi$$

define admissible parameters for small values of ε, as it is easily seen. With these parameters in (5.13), $J(\bar{R})$ becomes a function $f(\varepsilon)$ of ε, which has a minimum for $\varepsilon = 0$. The equation $f'(0) = 0$ gives the relation

$$\iint \{[v(E - G) - 2uF]\, \psi_u + [u(E - G) + 2vF]\, \psi_v\} \, du \, dv = 0. \quad (5.14)$$

Let now $\varphi(u, v)$ be a function having the properties required of $\psi(u, v)$ and having the additional property of vanishing on $u^2 + v^2 = 1$. The formulas[1]

$$\alpha = u + \varepsilon \varphi(u, v), \quad \beta = v$$

define then admissible parameters for small values of ε. Using these parameters in (5.13), we obtain

$$\iint [(E - G)\, \varphi_u + 2F \varphi_v] \, du \, dv = 0. \quad (5.15)$$

The equations (5.14) and (5.15) are both of the form

$$\iint [a(u, v)\, \lambda_u + b(u, v)\, \lambda_v] \, du \, dv = 0,$$

where $a(u, v)$, $b(u, v)$ have continuous first partial derivatives in $u^2 + v^2 < 1$. The function $\lambda(u, v)$ has to be sufficiently regular and

[1] We are using here again the method of the variation of the independent variables, which also played an important part in the non-parametric problem (see IV.12).

has to vanish on $u^2 + v^2 = 1$.[1] This is exactly the situation which arises in the discussion of the first variation of multiple integrals; the classical inference is that $a_u + b_v = 0$. Applying this to (5.14) and (5.15), we obtain[2]

$$\frac{\partial}{\partial u}[v(E-G) - 2uF] + \frac{\partial}{\partial v}[u(E-G) + 2vF] = 0\,, \qquad (5.16)$$

$$\frac{\partial}{\partial u}(E-G) + \frac{\partial}{\partial v}(2F) = 0\,, \qquad (5.17)$$

for $u^2 + v^2 < 1$. From (5.16) follows the existence, in $u^2 + v^2 < 1$, of a function $\Omega(u, v)$ such that

$$\Omega_v = v(E-G) - 2uF\,, \quad \Omega_u = -u(E-G) - 2vF\,. \qquad (5.18)$$

From (5.16), (5.17), (5.18) it follows by simple computations that $\Omega_{uu} + \Omega_{vv} = 0$, that is to say that Ω is harmonic. (5.14) can then be written in the form

$$\iint (\Omega_v \psi_u - \Omega_u \psi_v)\, du\, dv = 0\,,$$

where ψ is *not* supposed to vanish on $u^2 + v^2 = 1$. If we substitute for the harmonic function Ω its FOURIER expansion and if we put $\psi = r^n \cos n\Theta$ and $\psi = r^n \sin n\Theta$, where $r\cos\Theta = u$, $r\sin\Theta = v$, it follows immediately that Ω is constant. Hence $\Omega_u = 0$, $\Omega_v = 0$, and (5.18) gives the desired equations $E = G$, $F = 0$.

V.26. To apply the preceding result to the functional $A(T)$ of J. DOUGLAS, it is necessary to restate the problem $A(T) =$ minimum in the following form. Let Γ^* be a given JORDAN curve. Consider all surfaces S which admit of a representation

$$S: x = x(u, v)\,, \quad y = y(u, v)\,, \quad z = z(u, v)\,, \quad u^2 + v^2 \leqq 1\,,$$

with the following properties.

1. $x(u, v)$, $y(u, v)$, $z(u, v)$ have continuous partial derivatives of the first and second order in $u^2 + v^2 < 1$.

2. $x(u, v)$, $y(u, v)$, $z(u, v)$ are continuous in $u^2 + v^2 \leqq 1$, and the equations $x = x(u, v)$, $y = y(u, v)$, $z = z(u, v)$ define a monotonic transformation of $u^2 + v^2 = 1$ into a set on Γ^* (this implies that this set contains at least three distinct points, and on account of the assumed continuity it follows that the set covers the whole curve Γ^*).

We shall say that S is an *admissible surface, given in an admissible representation.* Let then the functional $J(S; R)$ be defined by

$$J(S; R) = \tfrac{1}{2}\iint (E + G)\,,$$

where R is the admissible representation which has been used to compute the integral.

The variation problem $J(S; R) =$ minimum is then clearly equivalent to the variation problem $A(T) =$ minimum of DOUGLAS.

---

[1] (5.14) holds even if λ does not vanish on the boundary.

[2] The lemma of HAAR (see IV.5) would permit to extend the proof to surfaces of class C'.

The application of the lemma in V.25 is now obvious. If R_0 is an admissible representation of an admissible surface S_0 such that $J(S_0;R_0)$ = minimum, that is to say such that $J(S_0; R_0) \leqq J(S; R)$ for all admissible representations of all the admissible surfaces, then in particular $J(S_0; R_0) \leqq J(S_0; R)$ for all the admissible representations R of S_0 itself. Hence R_0 is an isothermic representation, on account of the lemma in V.25.

Chapter VI.

The simultaneous problem in the parametric form. Generalizations.

VI.1. This problem has been investigated in the following statement. Given, in the xyz-space, a JORDAN curve Γ^*, consider all the continuous surfaces, of the type of the circular disc (see I.8), bounded by Γ^*, and suppose that the greatest lower bound $\mathfrak{a}(\Gamma^*)$ of their areas is finite. *Determine a solution* S *of problem* P_2 (see III.5), *such that* $\mathfrak{A}(S) = \mathfrak{a}(\Gamma^*)$.

While the method of GARNIER (see V.1) does not yield any information whatsoever as to the minimizing property of the solution obtained, both T. RADÓ and J. DOUGLAS proved that their respective methods yield a solution the area of which is a minimum. In either case, the proof depends essentially upon the existence theorem for conformal maps of polyhedrons, which has been the essential tool in the method reviewed in V.20 to V.23. These proofs will now be described briefly.

VI.2. Let $m(\Gamma^*)$ denote the smallest possible limit of the areas of sequences of polyhedrons $\mathfrak{P}_n$, such that the boundary polygon of $\mathfrak{P}_n$ converges, in the FRÉCHET sense, to Γ^*. This quantity $m(\Gamma^*)$ has been called by LEBESGUE, who introduced it in his Thesis[1], *the minimum area of* Γ^*. Supposing that $m(\Gamma^*)$ is finite, T. RADÓ[2] starts with the following *approximate problem*. Take three distinct points A, B, C on the unit circle $u^2 + v^2 = 1$, three distinct points A^*, B^*, C^* on Γ^*, and give an $\varepsilon > 0$. Determine three functions $x(u, v)$, $y(u, v)$, $z(u, v)$ with the following properties.

1. $x(u, v)$, $y(u, v)$, $z(u, v)$ are harmonic in $u^2 + v^2 < 1$, and
2. satisfy the inequalities

$$\iint\limits_{u^2+v^2<1} (E^{\frac{1}{2}} - G^{\frac{1}{2}})^2 \, du \, dv \leqq \varepsilon, \quad \iint\limits_{u^2+v^2<1} |F| \, du \, dv \leqq \varepsilon.$$

3. $x(u, v)$, $y(u, v)$, $z(u, v)$ are continuous in $u^2 + v^2 \leqq 1$, and the equations $x = x(u, v)$, $y = y(u, v)$, $z = z(u, v)$ carry $u^2 + v^2 = 1$ in a one-to-one and continuous way into a (not prescribed) JORDAN curve

[1] Intégrale, longueur, aire. Ann. Mat. pura appl. Vol. 7 (1902) pp. 231—359.

[2] The problem of the least area and the problem of PLATEAU. Math. Z. Vol. 32 (1930) pp. 763—796.

Γ^*_ε, the distance of which from Γ^* is $\leqq \varepsilon$. Furthermore, the points A, B, C are carried into three points $A^*_\varepsilon, B^*_\varepsilon, C^*_\varepsilon$ such that the distances $A^*A^*_\varepsilon$, $B^*B^*_\varepsilon$, $C^*C^*_\varepsilon$ are all three $\leqq \varepsilon$.

4. $\displaystyle\iint_{u^2+v^2<1} (EG - F^2)^{\frac{1}{2}}\, du\, dv \leqq m(\Gamma^*) + \varepsilon$.

The solution of this approximate problem is obtained by starting with a polyhedron $\mathfrak{P}$ with a boundary polygon $\mathfrak{p}$ such that both the difference $\mathfrak{A}(\mathfrak{P}) - m(\Gamma^*)$ and the distance $d(\mathfrak{p}, \Gamma^*)$ are less than a properly chosen $\sigma > 0$. If $x = \tilde{x}(u, v)$, $y = \tilde{y}(u, v)$, $z = \tilde{z}(u, v)$, $u^2 + v^2 \leqq 1$, is a properly normalized isothermic representation of $\mathfrak{P}$, then the harmonic functions $x(u, v)$, $y(u, v)$, $z(u, v)$, which coincide with $\tilde{x}(u, v)$, $\tilde{y}(u, v)$, $\tilde{z}(u, v)$ on $u^2 + v^2 = 1$, solve the approximate problem. The proof follows by a slight and obvious modification of the reasoning described in V.22.

VI.3. Keep then the points A, B, C, A^*, B^*, C^* fixed and solve the preceding approximate problem for $\varepsilon = \varepsilon_n$, $\varepsilon_n \to 0$. Denote by $x_n(u, v)$, $y_n(u, v)$, $z_n(u, v)$ the solution. A slight modification of the passage to the limit, used in V.23, shows that there exists a subsequence which converges to a solution of the exact problem, as stated in VI.1. Again, the approximate conditions

$$\iint (E_n^{\frac{1}{2}} - G_n^{\frac{1}{2}})^2 \leqq \varepsilon_n, \qquad \iint |F| \leqq \varepsilon_n,$$

prove to be just as effective as the exact conditions $E = G$, $F = 0$ would be. A few remarks are necessary in connection with the effect of condition 4 of the approximate problem. The immediate information, obtained from the passage to the limit, concerning the area $\mathfrak{A}(S)$ of the limit surface

$$S: x = x(u, v), \quad y = y(u, v), \quad z = z(u, v), \quad u^2 + v^2 \leqq 1$$

is that $\mathfrak{A}(S) = m(\Gamma^*)$. However, since S is bounded by Γ^*, we have $\mathfrak{A}(S) \leqq \mathfrak{a}(\Gamma^*)$ by the definition of $\mathfrak{a}(\Gamma^*)$, while, on the other hand, obviously $m(\Gamma^*) \leqq \mathfrak{a}(\Gamma^*)$. Consequently $\mathfrak{a}(\Gamma^*) = m(\Gamma^*) = \mathfrak{A}(S)$, that is to say the area of S is a minimum. It also follows that $\mathfrak{a}(\Gamma^*) = m(\Gamma^*)$, whenever $m(\Gamma^*)$ is finite. If $m(\Gamma^*) = +\infty$, then on account of $m(\Gamma^*) \leqq \mathfrak{a}(\Gamma^*)$ we also have $\mathfrak{a}(\Gamma^*) = +\infty$. Hence, for every Jordan curve Γ^*, we have $\mathfrak{a}(\Gamma^*) = m(\Gamma^*)$[1]. Thus the condition $m(\Gamma^*) < +\infty$, under which the approximate problem has been solved in VI.2, is equivalent to the condition $\mathfrak{a}(\Gamma^*) < +\infty$. It follows that *the simultaneous problem is solvable for every* Jordan *curve which bounds some continuous surface, of the type of the circular disc, with a finite area.* This condition obviously is satisfied if Γ^* is rectifiable. It is also clearly

[1] This generalizes a result of Lebesgue: Intégrale, longueur, aire. Ann. Mat. pura appl. Vol. 7 (1902) pp. 231–359.

satisfied if Γ^* is situated on a closed convex surface (since the area of such a surface is always finite and consequently the areas of the two regions, bounded on the surface by Γ^*, are also finite).

VI.4. Both T. RADÓ and J. DOUGLAS observe that for a general JORDAN curve Γ^* the condition $\mathfrak{a}(\Gamma^*) < +\infty$ is not satisfied. While this fact is rather obvious, the example given by J. DOUGLAS[1] is not quite convincing. J. DOUGLAS constructs a JORDAN curve Γ^* such that the measure of the orthogonal projection of Γ^* upon the xy-plane is $+\infty$, provided every point of the projection is counted with the proper multiplicity. From this it should follow that for every continuous surface S, of the type of the circular disc, the area $\mathfrak{A}(S)$ must also be $+\infty$, a conclusion which certainly is not obvious as it stands, since $\mathfrak{A}(S)$, in general, is *smaller* than the measure of the orthogonal projection of S upon a plane[2]. The reviewer takes the liberty to point out a simple way of obtaining more satisfactory examples. The orthogonal projection, upon the xy-plane, of a JORDAN curve Γ^* is a closed continuous curve C^*. Every point (x, y), not on C^*, has a definite integer $n(x, y)$ as its topological index with respect to C^*. Define $N(x, y)$ by $N(x, y) = |n(x, y)|$ if (x, y) is not on C^* and by $N(x, y) = 0$ if (x, y) is on C^*. Then $N(x, y)$ is a measurable function which is clearly zero outside of a sufficiently large circle K. Furthermore, $\mathfrak{A}(S) \geqq \iint N(x, y)\,dx\,dy$† for every continuous surface, of the type of the circular disc, bounded by Γ^*, and consequently $\mathfrak{a}(\Gamma^*) \geqq \iint N(x, y)\,dx\,dy$. It is clear on the other hand that if the projection C^* of Γ^* contains a properly arranged spiral-shaped arc, $\iint N(x, y)\,dx\,dy$ can be driven up to $+\infty$.

VI.5. To prove the minimizing property of the minimal surface obtained by his method, J. DOUGLAS[3] first observes that if m denotes the (by assumption finite) greatest lower bound of the functional $A(T)$ for the given JORDAN curve Γ^*, then $\mathfrak{A}(S) \geqq m$ for every continuous surface, of the type of the circle, bounded by Γ^*. Let us observe that this would be obvious if S would admit of an isothermic representation $\mathfrak{x} = \mathfrak{x}(u, v)$, $u^2 + v^2 \leqq 1$, $E = G$, $F = 0$. Indeed, denote by T the transformation of $u^2 + v^2 = 1$ into Γ^* defined by the equation of the surface and denote by $\tilde{x}(u, v)$, $\tilde{y}(u, v)$, $\tilde{z}(u, v)$ the harmonic functions coinciding with the components $x(u, v)$, $y(u, v)$, $z(u, v)$ of $\mathfrak{x}$ on $u^2+v^2=1$. Then

$$A(T) = \tfrac{1}{2}\iint (\tilde{E} + \tilde{G})\,.$$

---

[1] Solution of the problem of PLATEAU. Trans. Amer. Math. Soc. Vol. 33 (1931) pp. 320—321.

[2] See I.5 and I.6.

† This follows easily from the theorems in T. RADÓ: Über das Flächenmaß rektifizierbarer Flächen. Math. Ann. Vol. 100 (1928) pp. 445—479 § 2.

[3] Solution of the problem of PLATEAU. Trans. Amer. Math. Soc. Vol. 33 (1931) pp. 318—321. See also the abstract, Existence of a surface etc. Bull. Amer. Math. Soc. Vol. 36 (1930) pp. 796—797.

On account of the minimizing property of harmonic functions, we also have (cf. II.10)

$$\tfrac{1}{2}\iint(\tilde{E}+\tilde{G}) \leqq \tfrac{1}{2}\iint(E+G).$$

On account of $E = G$, $F = 0$, we have

$$\mathfrak{A}(S) = \tfrac{1}{2}\iint(E+G).$$

Finally $m \leqq A(T)$ by the definition of m. Consequently $m \leqq \mathfrak{A}(S)$.

It is then easy to understand the proof of J. DOUGLAS for a general S. The surface is approximated by polyhedrons and the inequality $m \leqq \mathfrak{A}(S)$ follows then, by several passages to the limit, from the fact that these approximating polyhedrons do admit of isothermic representations. While it follows from this reasoning that $m \leqq \mathfrak{a}(\Gamma^*)$, it is on the other hand obvious that we have, for the minimal surface S obtained by the method of DOUGLAS, $m = \mathfrak{A}(S)$, which implies $m \geqq \mathfrak{a}(\Gamma^*)$. Hence $\mathfrak{A}(S) = m = \mathfrak{a}(\Gamma^*)$. Thus it follows that S has the minimizing property.

It follows in this manner also that *the assumption $m < +\infty$, under which the method of* J. DOUGLAS *operates, is equivalent to the assumption that Γ^* bounds some continuous surface, of the type of the circular disc, with a finite area.*

It is strange that while the method of DOUGLAS is otherwise independent of the theory of conformal mapping, he needs this theory to determine the generality of the method and also to establish the fact that the solution has a minimum area. It should be observed however that J. DOUGLAS considers this part of his work as a stop-gap, and expects to develop a method entirely independent of the theory of conformal mapping[1].

VI.6. J. DOUGLAS gave another solution of the simultaneous problem (see VI.1) on the following basis[2]. Take a sequence of polyhedrons $\mathfrak{P}_n$, with boundary polygons converging to the given JORDAN curve Γ^*, such that $\mathfrak{A}(\mathfrak{P}_n)$ converges to the (by assumption finite) minimum area $m(\Gamma^*)$ of Γ^*. Let $\mathfrak{P}_n : \mathfrak{x} = \mathfrak{x}_n(u, v)$, $u^2 + v^2 \leqq 1$, be an isothermic representation of $\mathfrak{P}_n$. These equations define a transformation T_n of $u^2 + v^2 = 1$ into the boundary polygon $\mathfrak{p}_n$ of $\mathfrak{P}_n$, and if the isothermic representations $\mathfrak{x} = \mathfrak{x}_n(u, v)$ are properly normalized, then the sequence T_n will contain a subsequence converging to a transformation

$$T : x = \xi(\Theta), \quad y = \eta(\Theta), \quad z = \zeta(\Theta)$$

of the unit circle $u = \cos\Theta$, $v = \sin\Theta$ into Γ^*, such that the harmonic functions $x(u, v)$, $y(u, v)$, $z(u, v)$ with the boundary values $\xi(\Theta)$, $\eta(\Theta)$, $\zeta(\Theta)$ solve the simultaneous problem for Γ^*. Since the triples of harmonic

[1] See J. DOUGLAS: Solution of the problem of PLATEAU. Trans. Amer. Math. Soc. Vol. 33 (1931) p. 265.

[2] J. DOUGLAS: The mapping theorem of KOEBE and the problem of PLATEAU. J. Math. Physics, Massachusetts Inst. Technol. Vol. 10 (1931) pp. 106—130.

functions $\tilde{x}_n(u,v)$, $\tilde{y}_n(u,v)$, $\tilde{z}_n(u,v)$, which coincide on $u^2+v^2=1$ with the components $x_n(u,v)$, $y_n(u,v)$, $z_n(u,v)$ of $\mathfrak{x}_n(u,v)$, are obviously identical to the approximate solutions used in VI.2, this method differs, both in its general layout and in its details, very little from the one reviewed in VI.2 to VI.3.

VI.7. For JORDAN curves Γ^* such that $m(\Gamma^*)=+\infty$, J. DOUGLAS obtained, by a passage to the limit, the theorem that *problem* P_2 (see III.5) *admits of a solution S such that every interior portion of S has a minimum area with respect to its own boundary curve*[1]. The proof proceeds as follows. Take a sequence $\mathfrak{p}_n^*$ of simple closed polygons converging, in the sence of FRÉCHET, to Γ^*. For $\mathfrak{p}_n^*$, the simultaneous problem is certainly solvable; let

$$S_n: x = x_n(u,v), \quad y = y_n(u,v), \quad z = z_n(u,v), \quad u^2+v^2 \leqq 1, \tag{6.1}$$

be a solution, normalized by the condition that three distinct fixed points A, B, C on $u^2+v^2=1$ are carried into three distinct points A_n^*, B_n^*, C_n^* of $\mathfrak{p}_n^*$ which converge to three distinct points A^*, B^*, C^* on Γ^*. On account of the approximation theorem (see V.9), it is legitimate to suppose that the sequence converges, in $u^2+v^2\leqq 1$, to a solution

$$S: x = x(u,v), \quad y = y(u,v), \quad z = z(u,v), \quad u^2+v^2 \leqq 1$$

of problem P_2 for Γ^*. Since all the functions involved are harmonic and uniformly bounded, the convergence extends to partial derivatives and is uniform in every smaller concentric circle. Denote by $S^{(r)}$ the portion of S corresponding to $u^2+v^2\leqq r^2<1$. J. DOUGLAS proves then that $S^{(r)}$ has a minimum area, with respect to its own boundary curve, by computations whose geometrical background might be explained as follows. Denote by $l^{(r)}$ the length of the boundary curve of $S^{(r)}$, and let $S_n^{(r)}$, $l_n^{(r)}$ have similar meanings for S_n. Since all the partial derivatives converge uniformly in $u^2+v^2\leqq r^2<1$, it follows that $l_n^{(r)}\to l^{(r)}$, $\mathfrak{A}(S_n^{(r)})\to\mathfrak{A}(S^{(r)})$. It is then easily seen that we have a ring-shaped surface $\Sigma_n^{(r)}$, bounded by the boundary curves of $S_n^{(r)}$ and $S^{(r)}$, such that $\mathfrak{A}(\Sigma_n^{(r)})\to 0$.

Suppose there exists a continuous surface $\bar{S}^{(r)}$, of the type of the circular disc, with the same boundary as $S^{(r)}$ and such that $\mathfrak{A}(\bar{S}^{(r)})\leqq\mathfrak{A}(S^{(r)})-\varepsilon$, $\varepsilon>0$. Denote by $\bar{S}_n$ the surface $\bar{S}^{(r)}+\Sigma_n^{(r)}$. Then $\mathfrak{A}(\bar{S}_n^{(r)})\to\mathfrak{A}(\bar{S}^{(r)})$, and from the relations $\mathfrak{A}(\bar{S}^{(r)})\leqq\mathfrak{A}(S^{(r)})-\varepsilon$, $\mathfrak{A}(S_n^{(r)})\to\mathfrak{A}(S^{(r)})$ it follows that $\mathfrak{A}(\bar{S}_n^{(r)})<\mathfrak{A}(S_n^{(r)})$ for large values of n. Replacing then, for S_n, the portion $S_n^{(r)}$ by the surface $\bar{S}_n^{(r)}$, we obtain a surface with area less than S_n, in contradiction with the minimizing property of S_n.

[1] J. DOUGLAS: The least area property of the minimal surface determined by an arbitrary JORDAN contour. Proc. Nat. Acad. Sci. U. S. A. Vol. 17 (1931) pp. 211–216.

Thus $S^{(r)}$ has a minimum area with respect to its own boundary, and consequently every portion of $S^{(r)}$ has the same property. Since r was arbitrary, this proves the theorem.

VI.8. The reviewer has had the privilege of seeing the manuscript of a paper by J. E. McShane[1], in which the solution of the simultaneous problem is derived from general theorems, interesting in themselves.

The following definitions are used by McShane. A function $f(u, v)$, given in $u^2 + v^2 \leqq 1$, will be said to satisfy the *condition* (C) if

1. $f(u, v)$ is continuous in $u^2 + v^2 \leqq 1$, and

2. $f(u, v)$ is for almost every u an absolutely continuous function of v, and for almost every v an absolutely continuous function of u, and

3. the Dirichlet integral $\iint (f_u^2 + f_v^2)$, extended over $u^2 + v^2 < 1$, is finite.

Let again $f(u, v)$ be continuous in $u^2 + v^2 \leqq 1$. Let R be a subregion of $u^2 + v^2 \leqq 1$, and denote by $M(R)$, $m(R)$ the maximum and minimum of $f(u, v)$ in R, and by $M_b(R)$, $m_b(R)$ the maximum and minimum of $f(u, v)$ on the boundary of R. The least upper bound of $M(R) - M_b(R)$, $m_b(R) - m(R)$, for all subregions R of $u^2 + v^2 \leqq 1$, will be called the *monotonic deficiency* of $f(u, v)$ in $u^2 + v^2 \leqq 1$. Clearly, $f(u, v)$ is monotonic, is the sense explained in IV.18, if and only if the monotonic deficiency is zero.

VI.9. With these definitions we have the following *selection theorem*. Let $f_n(u, v)$ be a sequence of functions in $u^2 + v^2 \leqq 1$, such that 1. the Dirichlet integrals, extended over $u^2 + v^2 < 1$ are uniformly bounded, 2. every $f_n(u, v)$ satisfies condition (C), 3. the sequence converges uniformly on $u^2 + v^2 = 1$, and 4. the monotonic deficiency of $f_n(u, v)$ converges to zero for $n \to \infty$. *Then the sequence contains a subsequence which converges uniformly in* $u^2 + v^2 \leqq 1$, *and the limit function satisfies again condition* (C).

Except for the second half of the theorem, concerned with the condition (C), this theorem is a generalization of a similar selection theorem used by Lebesgue[2]. As observed by McShane, the proof of Lebesgue extends easily to this more general case. Condition (C) also can easily be handled on the basis of general theorems in the theory of functions of real variables.

VI.10. Another *selection theorem* used by McShane is concerned with *surfaces*. Let

$$S_n : x = x_n(u, v), \quad y = y_n(u, v), \quad z = z_n(u, v), \quad u^2 + v^2 \leqq 1 \qquad (6.2)$$

be a sequence of surfaces with the following properties.

[1] See the abstract: J. E. McShane: Parametrization of saddle-surfaces, with application to the problem of Plateau. Bull. Amer. Math. Soc. Vol. 38 (1932) pp. 810—811. The detailed presentation appears in the Trans. Amer. Math. Soc.

[2] Sur le problème de Dirichlet. Rend Circ. mat. Palermo Vol. 24 (1907) pp. 371—402.

1. All the coordinate functions satisfy condition (C) of VI.8.

2. The DIRICHLET integrals, over $u^2 + v^2 < 1$, of the coordinate functions are less than a fixed constant M independent of n.

3. The monotonic deficiency of $x_n(u, v)$, $y_n(u, v)$, $z_n(u, v)$ converges to zero.

4. The equations (6.2) take $u^2 + v^2 = 1$ in a one-to-one and continuous way into a JORDAN curve Γ_n^*. These curves Γ_n^* converge, in the sense of FRÉCHET, to a JORDAN curve Γ^*. There exist three distinct points A, B, C on $u^2 + v^2 = 1$ whose images A_n^*, B_n^*, C_n^* converge to three distinct points A^*, B^*, C^* of Γ^*.

Then the sequence (6.2) contains a subsequence which converges uniformly in $u^2 + v^2 \leqq 1$. The limit surface

$$S: x = x(u, v), \quad y = y(u, v), \quad z = z(u, v); \quad u^2 + v^2 \leqq 1, \tag{6.2.1}$$

is such that $x(u, v)$, $y(u, v)$, $z(u, v)$ are monotonic and satisfy condition (C). Furthermore, the equations (6.2.1) define, for $u^2 + v^2 = 1$, a continuous monotonic transformation of $u^2 + v^2 = 1$ into Γ^*, such that A, B, C are carried into A^*, B^*, C^*.

To prove this, denote by T_n the monotonic transformation of $u^2 + v^2 = 1$ into Γ_n^*, defined by (6.2) for $u^2 + v^2 = 1$. All the conditions for the selection theorem in I.22 being satisfied, there is a convergent subsequence, to be denoted again by T_n, which converges to a monotonic transformation T of $u^2 + v^2 = 1$ into Γ^*. The continuity of T follows then from assumption 2 by a reasoning similar to the lemma in V.15.

Now then, if a sequence of monotonic functions converges, on a closed interval, to a continuous function, then the convergence necessarily is uniform[1]. McSHANE observes that this obviously remains true for sequences of monotonic transformations. From the continuity of T it follows therefore that T_n converges uniformly to T; in other words, the sequence (6.2) contains a subsequence which converges uniformly on $u^2 + v^2 = 1$. The present selection theorem is then an immediate consequence of the selection theorem in VI.9.

VI.11. A continuous surface

$$S: x = x(u, v), \quad y = y(u, v), \quad z = z(u, v), \quad u^2 + v^2 \leqq 1 \tag{6.3}$$

is called by McSHANE a *saddle-surface* if $x(u, v)$, $y(u, v)$, $z(u, v)$ are monotonic. He proves that this property is independent of the particular choice of the parametric representation of the surface. His main theorem is then that *every saddle-surface which has a finite area and which is bounded by a* JORDAN *curve admits of isothermic parameters in the following generalized sense.*

[1] See for instance PÓLYA-SZEGÖ: Aufgaben und Lehrsätze Vol. 1 p. 63 problem 127.

If S is a saddle-surface, bounded by a JORDAN curve Γ^*, with finite area, then S admits of a representation (6.3) with the following properties.

1. $x(u, v)$, $y(u, v)$, $z(u, v)$ satisfy condition (C) of VI.8 and are monotonic functions.

2. $E = G$, $F = 0$ almost everywhere in $u^2 + v^2 < 1$.

The proof proceeds in the following steps. By the definition of $\mathfrak{A}(S)$, we have a sequence of polyhedrons $\mathfrak{P}_n$ such that $\mathfrak{P}_n \to S$ and $\mathfrak{A}(\mathfrak{P}_n) \to \mathfrak{A}(S)$. Choose three distinct fixed points A, B, C on $u^2 + v^2 = 1$, three distinct fixed points A^*, B^*, C^* on the boundary curve Γ^* of S, and three distinct points A_n^*, B_n^*, C_n^* on the boundary polygon of $\mathfrak{P}_n^*$, such that $A_n^* \to A^*$, $B_n^* \to B^*$, $C_n^* \to C^*$ (this is possible on account of $\mathfrak{P}_n \to S$). Let

$$\mathfrak{P}_n : x = x_n(u, v), \quad y = y_n(u, v), \quad z = z_n(u, v), \quad u^2 + v^2 \leqq 1 \tag{6.4}$$

be an isothermic representation of $\mathfrak{P}_n$ (see I.19), such that A, B, C are carried into A_n^*, B_n^*, C_n^*. Then, on account of $E_n = G_n$, $F_n = 0$ we have

$$\mathfrak{A}(\mathfrak{P}_n) = \tfrac{1}{2}\iint (E_n + G_n). \tag{6.5}$$

Since $\mathfrak{A}(\mathfrak{P}_n) \to \mathfrak{A}(S)$, and $\mathfrak{A}(S)$ is finite, the integrals (6.5) are uniformly bounded, and consequently the DIRICHLET integrals of $x_n(u, v)$, $y_n(u, v)$, $z_n(u, v)$ are less than a finite constant M independent of n. From $\mathfrak{P}_n \to S$ it follows, on account of the assumption that S is a saddle-surface, that the monotonic deficiencies of $x_n(u, v)$, $y_n(u, v)$, $z_n(u, v)$ converge to zero. Finally, the isothermic representation (6.4) of $\mathfrak{P}_n$ clearly satisfies condition (C) of VI.8. Thus all the assumptions of the selection theorem of VI.10 are satisfied, and therefore the sequence (6.4) contains a subsequence converging uniformly in $u^2 + v^2 \leqq 1$. For the sake of simplicity, let (6.4) denote such a subsequence. If $x^*(u, v)$, $y^*(u, v)$, $z^*(u, v)$ denote the limit functions, then

$$S : x = x^*(u, v), \quad y = y^*(u, v), \quad z = z^*(u, v), \quad u^2 + v^2 \leqq 1$$

is a representation[1] of S as required by the theorem. Indeed, 1. is satisfied on account of the selection theorem of VI.10. To verify 2., observe that on account of I.16, we have

$$\mathfrak{A}(S) = \iint (E^* G^* - F^{*2})^{\frac{1}{2}}.$$

On the other hand, from $x_n(u, v) \rightrightarrows x_n^*(u, v)$ it follows, on account of the lower semi-continuity of the DIRICHLET integral[2], that

$$\iint (x_u^{*2} + x_v^{*2}) \leqq \underline{\lim} \iint (x_{nu}^2 + x_{nv}^2).$$

Write the corresponding inequalities for the y and z coordinates and add. There follows

$$\tfrac{1}{2}\iint (E^* + G^*) \leqq \underline{\lim}\, \tfrac{1}{2}\iint (E_n + G_n) = \underline{\lim}\, \mathfrak{A}(\mathfrak{P}_n) = \mathfrak{A}(S),$$

since
$$\tfrac{1}{2}\iint (E_n + G_n) = \mathfrak{A}(\mathfrak{P}_n) \to \mathfrak{A}(S).$$

[1] In the sense of I.8. [2] See IV.20.

Thus we obtain the inequality

$$\tfrac{1}{2}\iint (E^* + G^*) \leqq \mathfrak{A}(S) = \iint (E^*G^* - F^{*2})^{\frac{1}{2}}, \tag{6.6}$$

while on the other hand

$$\iint (E^*G^* - F^{*2})^{\frac{1}{2}} \leqq \iint E^{*\frac{1}{2}} G^{*\frac{1}{2}} \leqq \tfrac{1}{2}\iint (E^* + G^*). \tag{6.7}$$

From (6.6) and (6.7) it follows immediately that $E^* = G^*$, $F^* = 0$ almost everywhere.

VI.12. Let

$$S: x = x(u, v), \quad y = y(u, v), \quad z = z(u, v), \quad u^2 + v^2 \leqq 1$$

be a continuous surface bounded by a JORDAN curve Γ^*, such that $x(u, v)$, $y(u, v)$, $z(u, v)$ satisfy condition (C). Then, on account of I.16, $\mathfrak{A}(S)$ is finite and is given by

$$\mathfrak{A}(S) = \iint (EG - F^2)^{\frac{1}{2}}.$$

McSHANE proves then that there exists a saddle-surface

$$S^*: x = x^*(u, v), \quad y = y^*(u, v), \quad z = z^*(u, v), \quad u^2 + v^2 \leqq 1,$$

with the following properties.

1. S^* is again bounded by Γ^*,
2. $\mathfrak{A}(S^*) \leqq \mathfrak{A}(S)$,
3. $x^*(u, v)$, $y^*(u, v)$, $z^*(u, v)$ satisfy condition (C) of VI.8. The construction of S^* is analogous to a construction used by LEBESGUE in the theory of the DIRICHLET problem[1]. Roughly speaking, if S is not itself a saddle-surface, then we shall have on S some closed plane curve $\overline{\Gamma}$ which bounds a portion $\overline{S}$ of S which is not in the plane of $\overline{\Gamma}$. If then we replace $\overline{S}$ by its orthogonal projection upon the plane of $\overline{\Gamma}$, we do not increase the area of S (condition (C) is necessary to justify this conclusion), and we bring S closer to being a saddle-surface. This operation, rigorously worded and applied a denumerable infinity of times, leads to the saddle-surface S^*.

VI.13. The preceding theorems lead to the following *selection theorem*. Let there be given a sequence

$$S_n: x = x_n(u, v), \quad y = y_n(u, v), \quad z = z_n(u, v), \quad u^2 + v^2 \leqq 1,$$

of saddle-surfaces, such that every S_n is bounded by a JORDAN curve Γ_n^*. Suppose that the areas $\mathfrak{A}(S_n)$ are uniformly bounded, and that Γ_n^* converges, in the FRÉCHET sense, to a JORDAN curve Γ^*.

Then the sequence S_n contains a subsequence which converges in the sense of FRÉCHET. The limit surface is a saddle-surface bounded by Γ^*.

[1] LEBESGUE: Sur le problème de DIRICHLET. Rend. Circ. mat. Palermo Vol. 24 (1907) pp. 371–402.

The proof follows immediately from VI.12 and from the selection theorem of VI.10.

VI.14. On the basis of the preceding theorems, McSHANE gives the following solution of the simultaneous problem (see VI.1). Denote by $\mathfrak{a}(\Gamma^*)$ the (by assumption finite) greatest lower bound of the areas of all continuous surfaces, of the type of the circular disc, bounded by Γ^*. Then first we have a sequence of polyhedrons $\mathfrak{P}_n$ such that $\mathfrak{A}(\mathfrak{P}_n) \to \mathfrak{a}(\Gamma^*)$ and such that the boundary polygon $\mathfrak{p}_n^*$ of $\mathfrak{P}_n$ converges, in the FRÉCHET sense, to Γ^*. The theorem of VI.12 clearly applies to polyhedrons, hence we can replace $\mathfrak{P}_n$ by a saddle-surface S_n, bounded by $\mathfrak{p}_n^*$, such that $\mathfrak{A}(S_n) \leqq \mathfrak{A}(\mathfrak{P}_n)$. Then clearly

$$\overline{\lim}\, \mathfrak{A}(S_n) \leqq \lim \mathfrak{A}(\mathfrak{P}_n) = \mathfrak{a}(\Gamma^*). \tag{6.8}$$

Thus the sequence $\mathfrak{A}(S_n)$ is bounded. Hence the selection theorem of VI.13 applies, and we obtain a surface S, limit of a subsequence of the sequence S_n, such that

1. S is bounded by Γ^*,
2. S is a saddle-surface,
3. $\mathfrak{A}(S) \leqq \underline{\lim}\, \mathfrak{A}(S_n)$, on account of the lower semi-continuity of the area.

From 3 and (6.8) it follows that $\mathfrak{A}(S) \leqq \mathfrak{a}(\Gamma^*)$, while from 1 it follows that $\mathfrak{a}(\Gamma^*) \leqq \mathfrak{A}(S)$. Hence $\mathfrak{A}(S) = \mathfrak{a}(\Gamma^*)$, that is to say S solves the least area problem for Γ^*. *While, in general, this fact alone would not permit any further conclusions* (see III.13), *the fact that S is a saddle-surface enables* McSHANE *to show that S is a minimal surface.* Indeed, S being a saddle-surface with a finite area, it admits of an isothermic representation in the sense of the theorem of VI.11. Let

$$S: x = x(u, v), \quad y = y(u, v), \quad z = z(u, v), \quad u^2 + v^2 \leqq 1 \tag{6.9}$$

be an isothermic representation of S, and denote by $\tilde{x}(u, v)$, $\tilde{y}(u, v)$, $\tilde{z}(u, v)$ the harmonic functions coinciding on $u^2 + v^2 = 1$ with $x(u, v)$, $y(u, v)$, $z(u, v)$. The reasoning of II.10 applies on account of condition (C) (see VI.8), and it follows that the area of the surface

$$\tilde{S}: x = \tilde{x}(u, v), \quad y = \tilde{y}(u, v), \quad z = \tilde{z}(u, v), \quad u^2 + v^2 \leqq 1$$

is less than the area of S, unless $x(u, v)$, $y(u, v)$, $z(u, v)$ are themselves harmonic functions. Hence it follows from the minimizing property of S that the functions $x(u, v)$, $y(u, v)$, $z(u, v)$ in (6.9) are harmonic. Consequently, the equations $E = G$, $F = 0$, which are known to be satisfied almost everywhere in $u^2 + v^2 < 1$, will hold everywhere, since E, F, G are continuous (even analytic). That is to say, (6.9) solves the simultaneous problem for Γ^*.

VI.15. A little discussion is however necessary to finish up the existence proof. The selection theorem of VI.10, which plays an important part in the preceding proof, yielded a limit surface

$$S: x = x(u, v), \quad y = y(u, v), \quad z = z(u, v), \quad u^2 + v^2 \leqq 1,$$

such that these equations define a continuous monotonic transformation T of $u^2 + v^2 = 1$ into the boundary curve Γ^*, three distinct points A, B, C given on $u^2 + v^2 = 1$ being carried into three distinct points A^*, B^*, C^* given on Γ^*. On the other hand, it does *not* follow from the proof of the selection theorem that T is a one-to-one transformation, and, as a consequence, the same remark applies to the solution of the simultaneous problem obtained by the method of McSHANE. This point can however be taken care of easily. If (6.9) represents the solution, then $x(u, v)$, $y(u, v)$, $z(u, v)$ are harmonic and $E = G$, $F = 0$ in $u^2 + v^2 < 1$. If then the equations do not carry $u^2 + v^2 = 1$ in a one-to-one way into Γ^*, then there exists (see I.23) an arc σ on $u^2 + v^2 = 1$ along which $x(u, v)$, $y(u, v)$, $z(u, v)$ all three reduce to constants. The reasoning at the end of V.10 shows then that $x(u, v)$, $y(u, v)$, $z(u, v)$ reduce to constants identically. This however contradicts the fact that there exists on $u^2 + v^2 = 1$ a triple of distinct points A, B, C which are carried by (6.9) into a triple of distinct points A^*, B^*, C^* of Γ^*.

It would be interesting to determine if, in the important theorem of VI.11 on the conformal maps of saddle-surfaces, the correspondence between the boundary curves is or is not necessarily one-to-one, although this mapping theorem is efficient enough as it stands, as far as the problem of PLATEAU is concerned.

VI.16. The existence of the solution of the simultaneous problem has several interesting implications, some of which will now be considered.

It has been observed (see III.14) that a minimal surface, bounded by a given curve, does not in general have a minimum area. H. A. SCHWARZ[1], by a discussion of the second variation, obtained conditions under which a given piece S of a minimal surface has a minimum area with respect to surfaces bounded by the same curve and sufficiently close to S (relative minimum). The recent results permit the establishing of the following theorem concerned with an absolute minimum[2].

If the boundary curve Γ^* of a minimal surface S has a simply-covered convex curve as its parallel or central projection upon some plane, then the area of S is a minimum with respect to all surfaces bounded by Γ^* (surface means continuous surface of the type of the circular disc).

First, on account of the remark at the end of VI.3, the minimum area of Γ^* is finite, and hence the simultaneous problem has a solution S^* (see VI.3). On account of the uniqueness theorem of III.11, S^* is identical to S, which proves the theorem.

VI.17. Let S denote a regular minimal surface (see III.4). A sufficiently small vicinity S_0 of any point P_0 of S has then a simply covered

[1] Gesammelte Mathematische Abhandlungen Vol. 1 pp. 151—167 and pp. 224 to 269.

[2] T. RADÓ: Contributions to the theory of minimal surfaces. Acta Litt. Sci. Szeged Vol. 6 (1932) pp. 1—20.

orthogonal projection upon, say, the xy-plane. Hence P_0 is contained in a portion S_0^* of S_0 such that the xy-projection of the boundary curve of S_0^* is a convex curve (a circle, if we wish). Hence, on account of VI.16, *the area of a minimal surface, in the sense of differential geometry, is an absolute minimum in the small*[1].

VI.18. Let us consider now the non-parametric problem (Chapter IV). Denote by C a convex JORDAN curve in the xy-plane, and by $\varphi(P)$ a continuous function of the point P varying on C. The equation $z = \varphi(P)$ determines, in the xyz-space, a JORDAN curve Γ^*, and C is the simply-covered orthogonal projection of Γ^*. Since C is convex, the simultaneous problem has a solution S for Γ^* (see the end of VI.3), and on account of III.10, S can be represented by an equation $z = z(x, y)$, where $z(x, y)$ is continuous in and on C, analytic in C, equal to $\varphi(P)$ on C, and satisfies in C the partial differential equation

$$(1 + q^2)r - 2pqs + (1 + p^2)t = 0.$$

That is to say, *the boundary value problem is solvable for every continuous boundary function $\varphi(P)$ given on a convex curve C.*[2] The three-point condition (see IV.2) is thus seen to be superfluous.

It also follows that the solution $z = z(x, y)$ defines a minimal surface whose area is a minimum with respect to all continuous surfaces, of the type of the circular disc, bounded by the same curve (and not only with respect to surfaces $z = z(x, y)$, where $z(x, y)$ satisfies the LIPSCHITZ condition, as it would follow from the existence proof described in IV.17 to IV.26). Thus $z(x, y)$ solves the problem $\mathfrak{A}(z) =$ minimum to be considered in the next section VI.19.

VI.19. Let C and $\varphi(P)$ have the same meaning as in the preceding section VI.18. Consider the totality of all those functions $z(x, y)$ which are continuous in and on C and are equal to $\varphi(P)$ on C. Denote by $\mathfrak{A}(z)$ the area, in the sense of LEBESGUE, of the surface $z = z(x, y)$. Consider the problem $\mathfrak{A}(z) =$ minimum. On account of VI.18, the problem has a solution $z_1(x, y)$ (since C is supposed to be convex). McSHANE proved that the solution is *unique*[3]. To see this, observe first that

$$a = \min \mathfrak{A}(z)$$

is finite on account of the convexity of C (see the end of VI.3). Suppose we have another solution z_2. Then

$$\mathfrak{A}(z_1) = \mathfrak{A}(z_2) = a. \tag{6.10}$$

[1] T. RADÓ: Contributions to the theory of minimal surfaces. Acta Litt. Sci. Szeged Vol. 6 (1932) pp. 1–20.

[2] T. RADÓ: The problem of the least area and the problem of PLATEAU. Math. Z. Vol. 32 (1930) pp. 795–796. The theorem has been stated, without proof, by S. BERNSTEIN: Sur les équations du Calcul des Variations. Ann. École norm. Vol. 29 (1912) pp. 431–485. See in particular pp. 484–485.

[3] McSHANE: On a certain inequality of STEINER. Ann. of Math. Vol. 33 (1932) pp. 123–138.

Clearly, $\frac{1}{2}(z_1 + z_2)$ also reduces to $\varphi(P)$ on C. Hence

$$\mathfrak{A}\left(\frac{z_1+z_2}{2}\right) \geqq a\,,$$

while the inequality of STEINER (see I.17) gives

$$\mathfrak{A}\left(\frac{z_1+z_2}{2}\right) \leqq \tfrac{1}{2}\left(\mathfrak{A}(z_1) + \mathfrak{A}(z_2)\right) = a\,.$$

Thus

$$\mathfrak{A}\left(\frac{z_1+z_2}{2}\right) = \tfrac{1}{2}\left(\mathfrak{A}(z_1) + \mathfrak{A}(z_2)\right). \tag{6.11}$$

Consequently, as shown by McSHANE (see I.17), $p_1 = p_2$, $q_1 = q_2$ almost everywhere (p, q denote the partial derivatives of the first order). Well then, z_1 is analytic and thus

$$\mathfrak{A}(z_1) = \iint (1 + p_1^2 + q_1^2)^{\frac{1}{2}}\,. \tag{6.12}$$

Since $p_1 = p_2$, $q_1 = q_2$ almost everywhere, we have

$$\iint (1 + p_1^2 + q_1^2)^{\frac{1}{2}} = \iint (1 + p_2^2 + q_2^2)^{\frac{1}{2}}\,. \tag{6.13}$$

From (6.10), (6.12) and (6.13) it follows that

$$\mathfrak{A}(z_2) = \iint (1 + p_2^2 + q_2^2)^{\frac{1}{2}}\,.$$

Thus $\mathfrak{A}(z_1)$ and $\mathfrak{A}(z_2)$ are both given by the classical integral formula. In this case however, as shown by McSHANE (see I.17), (6.11) implies that $z_2 - z_1$ is constant. Since $z_2 - z_1 = 0$ on C, it follows that $z_2 \equiv z_1$.

VI.20. McSHANE observed[1] that the preceding result yields the following generalization of the theorem of IV.10.

Let the continuous surface S be a solution of the least area problem for a JORDAN *curve Γ^*. Suppose that every point P_0 of S is comprised in a portion S_0 of S which has a simply-covered orthogonal projection upon some plane. Then S is analytic (and is a minimal surface in the sense of differential geometry).*

To prove this, consider any one of the portions S_0 described in the theorem, and suppose that S_0 has a simply covered xy-projection for instance. Denote by C any convex JORDAN curve, comprised with its interior in the projection of S_0. Denote by S_0^* the portion of S_0 which is projected into the JORDAN region bounded by C. Let $z = z_0^*(x, y)$ be the equation of S_0^* and denote by $\varphi_0^*(P)$ the function to which $z_0^*(x, y)$ reduces on C. Then $z_0^*(x, y)$ clearly solves the problem $\mathfrak{A}(z) =$ minimum for the convex JORDAN curve C and the continuous boundary function $\varphi_0^*(P)$. Since the solution of the problem is unique (see VI.19), $z_0^*(x, y)$ coincides with the analytic solution whose existence is secured by VI.18.[2] This proves the theorem.

In IV.10, S has been supposed to be a regular surface of class C'. This assumption implies that S has, in the small, a simply covered

[1] McSHANE: On a certain inequality of STEINER. Ann. of Math. Vol. 33 (1932) pp. 123—138.

[2] The existence theorem of IV.16 is not sufficiently general for the present application, on account of the three-point condition.

projection upon a properly chosen plane. The result of McSHANE shows that this property is sufficient in itself to secure the analytic character of S. Thus the analytic character of a solution of the least area problem depends upon *regularity in the topological sense*, rather than upon regularity with respect to differential coefficients. It would be interesting to decide if it is sufficient to suppose that S is, in the small, a JORDAN surface (one-to-one and continuous image of the circular disc).

VI.21. Up to this time, we only considered surfaces of the type of the circular disc, bounded by a given JORDAN curve. All the problems we discussed so far could be stated for surfaces of any given topological type, bounded by any number of given curves. We conclude this report by a brief review of certain results due to J. DOUGLAS concerning problems of this generalized type[1].

Let there be given two non-intersecting JORDAN curves Γ_1, Γ_2 in the xyz-space. Determine in the uv-plane a circular ring bounded by two circles C_1, C_2 with center at the origin and with radii R_1 and $R_2 > R_1$, and three functions $x(u,v)$, $y(u,v)$, $z(u,v)$ with the following properties[2].

1. $x(u,v)$, $y(u,v)$, $z(u,v)$ are harmonic for $R_1^2 < u^2 + v^2 < R_2^2$,
2. and satisfy there the equations $E = G$, $F = 0$, and
3. $x(u,v)$, $y(u,v)$, $z(u,v)$ are continuous in $R_1^2 \leqq u^2 + v^2 \leqq R_2^2$ and the equations $x = x(u,v)$, $y = y(u,v)$, $z = z(u,v)$ carry the circles C_1 and C_2 in a one-to-one and continuous way into the given JORDAN curves Γ_1 and Γ_2 respectively.

VI.22. In case Γ_1, Γ_2 are both in the xy-plane, and for instance Γ_1 is enclosed by Γ_2, then the above problem requires to map the ring-shaped region bounded by Γ_1 and Γ_2 upon a circular ring in a one-to-one and continuous and in the interior conformal way, as is easily seen by a reasoning similar to the one used in V.19. As is well known in this special case, the circular ring cannot be given arbitrarily, the ratio of the radii of the (concentric) boundary circles being invariable under conformal mapping. Therefore, in the problem stated in VI.21, the ratio $q = R_1/R_2$ is to be considered as unknown.

VI.23. J. DOUGLAS[3] treats the problem of VI.21 by generalizing his method for the one-contour case. Take a q satisfying $0 < q < 1$, and

---

[1] J. DOUGLAS announced [see J. DOUGLAS: Solution of the problem of PLATEAU. Trans. Amer. Math. Soc. Vol. 33 (1931) p. 264] that his methods are adequate for the solution of the most general problem (any number of boundary curves, minimal surface of any prescribed topological type).

[2] J. DOUGLAS considers the n-dimensional problem (in the sense explained in V.19) and points out that the value of n does not make any difference. We restrict ourselves therefore to the case $n = 3$.

[3] The problem of PLATEAU for two contours. J. Math. Physics, Massachusetts Inst. Technol. Vol. 10 (1931) pp. 315–359.

two concentric circles C_1, C_2 with center at $u = 0$, $v = 0$ and with radii R_1, R_2 such that $q = R_1/R_2$. Denote by

$$T_1: x = \xi_1(\Theta), \quad y = \eta_1(\Theta), \quad z = \zeta_1(\Theta),$$
$$T_2: x = \xi_2(\Theta), \quad y = \eta_2(\Theta), \quad z = \zeta_2(\Theta),$$

monotonic transformations of the circles C_1, C_2 into Γ_1, Γ_2 respectively. Denote by $x_1(u, v)$, $y_1(u, v)$, $z_1(u, v)$ the harmonic functions, defined in $u^2 + v^2 < R_1^2$ and obtained by means of the POISSON integral formula, using $\xi_1(\Theta)$, $\eta_1(\Theta)$, $\zeta_1(\Theta)$ as boundary functions. Let $x_2(u, v)$, $y_2(u, v)$, $z_2(u, v)$ have a similar meaning for $u^2 + v^2 < R_2^2$ and $x_{12}(u, v)$, $y_{12}(u, v)$, $z_{12}(u, v)$ for $R_1^2 < u^2 + v^2 < R_2^2$. Put

$$A(T_1) = \tfrac{1}{2}\iint (E_1 + G_1), \quad u^2 + v^2 < R_1^2, \tag{6.14}$$

$$A(T_2) = \tfrac{1}{2}\iint (E_2 + G_2), \quad u^2 + v^2 < R_2^2, \tag{6.15}$$

$$A(T_1, T_2; q) = \tfrac{1}{2}\iint (E_{12} + G_{12}), \quad R_1^2 < u^2 + v^2 < R_2^2. \tag{6.16}$$

We use again E, F, G, with the proper subscripts, to denote the first fundamental quantities of a surface.

VI.24. Denote by $m(\Gamma_1)$, $m(\Gamma_2)$, $m(\Gamma_1, \Gamma_2)$ the greatest lower bounds of $A(T_1)$, $A(T_2)$, $A(T_1, T_2; q)$ for all possible monotonic transformations T_1, T_2 and for all values of q, $0 < q < 1$. Suppose these lower bounds are finite (the geometrical interpretation of this condition will be considered in VI.29).

It follows from the definitions that

$$m(\Gamma_1, \Gamma_2) \leqq m(\Gamma_1) + m(\Gamma_2).$$

J. DOUGLAS proves that *if*

$$m(\Gamma_1, \Gamma_2) < m(\Gamma_1) + m(\Gamma_2),$$

then the problem in VI.21 *is solvable.* The existence proof is similar to the one he developed for the one-contour case, although a great deal more involved. The proof is based on a detailed study of $A(T_1, T_2; q)$ in its dependence upon T_1, T_2, q. We are going to indicate the main steps.

VI.25. Develop the functions $\xi_1(\Theta)$, $\eta_1(\Theta)$, $\zeta_1(\Theta)$, which define the transformation T_1, into a FOURIER series. Denote by a_{1m}, b_{1m} the vectors whose components are the coefficients of $\cos m\Theta$, $\sin m\Theta$ respectively in the developments of $\xi_1(\Theta)$, $\eta_1(\Theta)$, $\zeta_1(\Theta)$. Let a_{2m}, b_{2m} have the same meaning with respect to T_2. Then DOUGLAS finds that

$$A(T_1, T_2; q) = \frac{4\pi\alpha^2}{-\log q} + \frac{\pi}{2}\sum_{m=1}^{\infty} m\,\frac{(1+q^{2m})(a_{1m}^2 + b_{1m}^2 + a_{2m}^2 + b_{2m}^2) - 4q^m(a_{1m}a_{2m} + b_{1m}b_{2m})}{1-q^{2m}},$$

where

$$\alpha = \tfrac{1}{2}(a_{10} - a_{20}).$$

It is necessary for the sequel to control $A(T_1, T_2; q)$ in the vicinity of $q = 0$ and $q = 1$. DOUGLAS finds, for q close to zero, the development

$$A(T_1, T_2; q) = A(T_1) + A(T_2) + \frac{4\pi\alpha^2}{-\log q} + q \times (\text{power series of } q). \quad (6.17)$$

For q close to 1, the following asymptotic expression is obtained:

$$A(T_1, T_2; q) \sim \frac{3\pi\alpha^2 + \frac{1}{2}\int_0^{2\pi} p(\Theta)\, d\Theta}{1-q} \quad (6.18)$$

and

$$\frac{dA(T_1, T_2; q)}{dq} \sim \frac{3\pi\alpha^2 + \frac{1}{2}\int_0^{2\pi} p(\Theta)\, d\Theta}{(1-q)^2}, \quad (6.19)$$

where

$$p(\Theta) = (\xi_1(\Theta) - \xi_2(\Theta))^2 + (\eta_1(\Theta) - \eta_2(\Theta))^2 + (\zeta_1(\Theta) - \zeta_2(\Theta))^2.$$

The numerators in (6.18) and (6.19) are clearly $\geqq \pi d^2$, where d is the by assumption positive shortest distance of Γ_1 and Γ_2. DOUGLAS also observes that the approximations (6.18) and (6.19) are uniform, provided Γ_1 and Γ_2 are restricted to fixed bounded regions.

VI.26. By the definition of $m(\Gamma_1, \Gamma_2)$, there exists a sequence $(T_1^{(n)}, T_2^{(n)}; q^{(n)})$ such that $A(T_1^{(n)}, T_2^{(n)}; q^{(n)}) \to m(\Gamma_1, \Gamma_2)$. For a properly chosen subsequence, which we denote again by $(T_1^{(n)}, T_2^{(n)}; q^{(n)})$, $T_1^{(n)}$ and $T_2^{(n)}$ will converge toward transformations T_1, T_2 and $q^{(n)}$ will converge to a limit q. In the present case it is impossible to normalize the sequences $T_1^{(n)}$, $T_2^{(n)}$ as it has been done in the one-contour case by prescribing the images of several points, since the conformal maps of a circular ring upon itself are determined as soon as the image of a single boundary point is known. Consequently, the selection theorem of I.22 cannot be referred to directly. An easy discussion shows however that the limit transformation T_1 is either a monotonic transformation in the sense of I.21 or otherwise T_1 is *degenerate* in the sense that it carries the whole circle into a single point. The same holds for the limit transformation T_2.

The first thing to do is therefore to show that T_1, T_2 are non-degenerate and that q is different from 0 and from 1. Otherwise $(T_1, T_2; q)$ could not serve to solve the problem $A(T_1, T_2; q) = \text{minimum}$. All these possibilities are excluded, as DOUGLAS finds, by the assumptions $m(\Gamma_1, \Gamma_2) < m(\Gamma_1) + m(\Gamma_2)$ and $m(\Gamma_1, \Gamma_2) < \infty$. It follows for instance from (6.17) for $q^{(n)} \to 0$ that

$$\lim A(T_1^{(n)}, T_2^{(n)}; q^{(n)}) \geqq m(\Gamma_1) + m(\Gamma_2),$$

and hence, since
$$\lim A(T_1^{(n)}, T_2^{(n)}; q^{(n)}) = m(\Gamma_1, \Gamma_2),$$

that $m(\Gamma_1, \Gamma_2) \geqq m(\Gamma_1) + m(\Gamma_2)$. For $q^{(n)} \to 1$ it follows from (6.18) that

$$A(T_1^{(n)}, T_2^{(n)}; q^{(n)}) \to +\infty.$$

VI.27. Thus we have a q with $0<q<1$, a circular ring $R_1^2 \leq u^2+v^2 \leq R_2^2$ with $R_1/R_2 = q$, and monotonic transformations

$$T_1: x = \xi_1(\Theta), \quad y = \eta_1(\Theta), \quad z = \zeta_1(\Theta),$$
$$T_2: x = \xi_2(\Theta), \quad y = \eta_2(\Theta), \quad z = \zeta_2(\Theta)$$

of the boundary circles C_1, C_2 of the ring into the given JORDAN curves Γ_1, Γ_2, such that

$$A(T_1, T_2; q) = m(\Gamma_1, \Gamma_2).$$

From $m(\Gamma_1, \Gamma_2) < +\infty$ it follows again, by means of the expression (6.16) and, for instance, the lemma in V.15, that T_1, T_2 are both continuous.

VI.28. Denote by $x_{12}(u, v)$, $y_{12}(u, v)$, $z_{12}(u, v)$ the harmonic functions, defined in $R_1^2 \leq u^2 + v^2 \leq R_2^2$, which reduce to $\xi_1(\Theta)$, $\eta_1(\Theta)$, $\zeta_1(\Theta)$ on C_1 and to $\xi_2(\Theta)$, $\eta_2(\Theta)$, $\zeta_2(\Theta)$ on C_2. DOUGLAS shows that these functions satisfy the equations $E_{12} = G_{12}$, $F_{12} = 0$, by computing the first variation of the functional $A(T_1, T_2; q)$. The variations of T_1, T_2 are controlled by computations similar to those in V.18. In addition variations of q must also be considered in the present case. These are handled by means of the asymptotic expression (6.19). Thus it follows that $x_{12}(u, v)$, $y_{12}(u, v)$, $z_{12}(u, v)$ satisfy all the conditions of the problem, except possibly for condition 3. Now that $E_{12} = G_{12}$, $F_{12} = 0$ has already been shown, this last point can be taken care of easily. If the correspondence between the boundaries would not be one-to-one, then x_{12}, y_{12}, z_{12} would reduce to constants on a whole arc of the boundary of the circular ring, and consequently (cf. the end of V.10) they would reduce to constants identically. Then T_1, T_2 would necessarily both be degenerate, and this possibility has been excluded previously.

VI.29. There remains to discuss the *geometrical meaning of* the condition $m(\Gamma_1, \Gamma_2) < m(\Gamma_1) + m(\Gamma_2)$. Denote by $\mathfrak{P}_{12}^{(n)}$ a sequence of polyhedrons, of the topological type of the circular ring, such that the boundary polygons of $\mathfrak{P}_{12}^{(n)}$ converge, in the FRÉCHET sense, to Γ_1 and Γ_2 respectively. Consider $\underline{\lim}\, \mathfrak{A}(\mathfrak{P}_{12}^{(n)})$. Then DOUGLAS observes that the minimum of $\underline{\lim}\, \mathfrak{A}(\mathfrak{P}_{12}^{(n)})$, for all possible sequences $\mathfrak{P}_{12}^{(n)}$ with the properties required above, is equal to $m(\Gamma_1, \Gamma_2)$. Similarly, $m(\Gamma_1)$ and $m(\Gamma_2)$ are equal to the minimum areas of the curves Γ_1, Γ_2 in the LEBESGUE sense. The area of the minimal surface S whose existence has been proved in VI.23 to VI.28 is clearly equal to $m(\Gamma_1, \Gamma_2)$. Since obviously the area of any continuous surface, of the type of a circular ring, bounded by Γ_1 and Γ_2 is at least equal to $m(\Gamma_1, \Gamma_2)$, it follows that the area of the minimal surface S is a minimum with respect to all these surfaces. Thus it follows that *the condition* $m(\Gamma_1, \Gamma_2) < m(\Gamma_1) + m(\Gamma_2)$ *has the following geometrical meaning: the minimum of the areas of all continuous surfaces, of the type of the circular ring, bounded by* Γ_1 *and* Γ_2 *is less than the sum* *of the minima of the areas of all continuous surfaces, of the type of the circular disc, bounded by Γ_1 and Γ_2 separately.*

J. DOUGLAS quotes several cases in which the condition $m(\Gamma_1, \Gamma_2) < m(\Gamma_1) + m(\Gamma_2)$ is satisfied. An interesting case is that of two interlacing JORDAN curves. Another important case is that of two JORDAN curves Γ_1, Γ_2 in the same plane, such that Γ_2 for instance encloses Γ_1. In this case, the problem of VI.21 reduces to map the ring-shaped region between Γ_1 and Γ_2 in a one-to-one and continuous and in the interior conformal way upon a circular ring. J. DOUGLAS emphasizes the fact that his method yields a new proof for the existence of this map.

J. DOUGLAS observes that in order to obtain the geometrical meaning of $m(\Gamma_1, \Gamma_2)$, $m(\Gamma_1)$, $m(\Gamma_2)$ described above, it is necessary to use the existence of conformal maps of polyhedrons, in a similar way as in the one-contour case (see VI.5). He considers this again as a stop-gap and is planning to develop a method entirely independent of the theory of conformal mapping.

VI.30. In what precedes it has been supposed that $m(\Gamma_1, \Gamma_2)$ is finite. DOUGLAS generalizes his result in the following way. Denote by $e(\Gamma_1, \Gamma_2)$ the maximum of $\overline{\lim}\,[m(\Gamma_1^{(n)}) + m(\Gamma_2^{(n)}) - m(\Gamma_1^{(n)}, \Gamma_2^{(n)})]$ for all sequences $\Gamma_1^{(n)}, \Gamma_2^{(n)}$ such that $\Gamma_1^{(n)} \to \Gamma_1$, $\Gamma_2^{(n)} \to \Gamma_2$ and such that $m(\Gamma_1^{(n)}, \Gamma_2^{(n)})$ is finite. These conditions are certainly satisfied by polygons $\Gamma_1^{(n)}, \Gamma_2^{(n)}$, for instance. DOUGLAS obtains then the theorem that *the problem in* VI.21 *is solvable for Γ_1, Γ_2 if $e(\Gamma_1, \Gamma_2) > 0$.*

The method consists of solving the problem for a sequence $\Gamma_1^{(n)}, \Gamma_2^{(n)}$ such that $m(\Gamma_1^{(n)}) + m(\Gamma_2^{(n)}) - m(\Gamma_1^{(n)}, \Gamma_2^{(n)}) \to e(\Gamma_1, \Gamma_2)$ and passing then to the limit. The passage to the limit is substantially the same as that described in VI.26, and is controlled by the fact that the approximations (6.18) and (6.19) are uniform provided the curves involved are restricted to fixed finite regions of the space.

The condition $e(\Gamma_1, \Gamma_2) > 0$ is satisfied, as DOUGLAS observes, for *any two interlacing* JORDAN *curves.*

VI.31. Closely related to the two-contour case is the problem, considered by DOUGLAS, to *determine a minimal surface of the topological type of the* MÖBIUS *strip and bounded by a given* JORDAN *curve*[1]. The analytic formulation of the problem is as follows[2].

Given a JORDAN curve Γ, determine a q satisfying $0 < q < 1$ and three functions $x(u,v)$, $y(u,v)$, $z(u,v)$ with the following properties.

1. $x(u,v)$, $y(u,v)$, $z(u,v)$ are harmonic for $q^2 < u^2 + v^2 < 1$ and
2. satisfy there the equations $E = G$, $F = 0$.

---

[1] J. DOUGLAS: One-sided minimal surfaces with a given boundary. Trans. Amer. Math. Soc. Vol. 34 (1932) pp. 731–756.

[2] J. DOUGLAS considers the n-dimensional problem (in the sense explained in V.19) and points out that the value of n does not make any difference. We restrict ourselves therefore to the case $n = 3$.

3. $x(u,v)$, $y(u,v)$, $z(u,v)$ are continuous in $q^2 \leqq u^2 + v^2 \leqq 1$ and the equations $x = x(u,v)$, $y = y(u,v)$, $z = z(u,v)$ carry the circles $u^2 + v^2 = 1$ and $u^2 + v^2 = q^2$ in a one-to-one and continuous way into the given JORDAN curve Γ.

4. If $\xi(\Theta)$, $\eta(\Theta)$, $\zeta(\Theta)$ are the boundary values of $x(u,v)$, $y(u,v)$, $z(u,v)$ on $u^2 + v^2 = 1$, then the boundary values on $u^2 + v^2 = q^2$ are $\xi(\Theta + \pi)$, $\eta(\Theta + \pi)$, $\zeta(\Theta + \pi)$.

VI.32. Condition 4 implies that the equations $x = x(u,v)$, $y = y(u,v)$ $z = z(u,v)$ define a minimal surface of the type of the MÖBIUS strip. Indeed, if (u,v) is a point of $q^2 \leqq u^2 + v^2 \leqq 1$, then let (u^*, v^*) denote the point obtained by reflecting (u,v) on the circle $u^2 + v^2 = q$ and then rotating the resulting point around $(0,0)$ by the angle π. The points (u,v), (u^*, v^*) will be called *elliptically inverse* to each other with respect to the circle $u^2 + v^2 = q$. Condition 4 expresses then that

$$x(u,v) = x(u^*, v^*), \quad y(u,v) = y(u^*, v^*), \quad z(u,v) = z(u^*, v^*) \quad (6.20)$$

on the boundary of the ring $q^2 \leqq u^2 + v^2 \leqq 1$. Consequently (6.20) also holds in the whole ring, since elliptic inversion transforms harmonic functions into harmonic functions, and since harmonic functions are determined by their boundary values. From (6.20) it follows that the equations $x = x(u,v)$, $y = y(u,v)$, $z = z(u,v)$ determine a minimal surface of the type of the MÖBIUS strip (every point of which is obtained twice).

VI.33. Denote by

$$T: x = \xi(\Theta), \quad y = \eta(\Theta), \quad z = \zeta(\Theta),$$

a monotonic transformation of $u = \cos\Theta$, $v = \sin\Theta$ into the given JORDAN curve Γ, and denote by T^* the transformation

$$T^*: x = \xi(\Theta + \pi), \quad y = \eta(\Theta + \pi), \quad z = \zeta(\Theta + \pi).$$

Denote by q a number satisfying $0 < q < 1$. Let $x(u,v)$, $y(u,v)$, $z(u,v)$ denote the harmonic functions, defined in $q^2 \leqq u^2 + v^2 \leqq 1$ by means of the POISSON integral formula, using as boundary functions $\xi(\Theta)$, $\eta(\Theta)$, $\zeta(\Theta)$ on $u^2 + v^2 = 1$ and $\xi(\Theta + \pi)$, $\eta(\Theta + \pi)$, $\zeta(\Theta + \pi)$ on $u^2 + v^2 = q^2$. Put

$$A(T;q) = \tfrac{1}{2} \cdot \tfrac{1}{2} \iint (E + G),$$

the integration being extended over $q^2 < u^2 + v^2 < 1$. The method of DOUGLAS consists then in minimizing $A(T;q)$; the harmonic functions $x(u,v)$, $y(u,v)$, $z(u,v)$ corresponding to the minimizing set $(T;q)$ solve then the problem in VI.31. DOUGLAS observes that with the notations of VI.23 we can write

$$A(T;q) = \tfrac{1}{2} A(T, T^*; q).$$

Thus the present problem appears as a *limit case of the two-contour problem*. As a matter of fact, the treatment of the present problem follows step by step the treatment of the two-contour case, described in VI.23 to VI.30. We restrict ourselves therefore to state the result.

Denote by $\overline{m}(\Gamma)$ the greatest lower bound of $A(T; q)$ and let $m(\Gamma)$ have the same meaning with respect to Γ as in the one contour case. Then it is easily seen that $\overline{m}(\Gamma) \leq m(\Gamma)$. *Whenever* $\overline{m}(\Gamma) < m(\Gamma)$, *the curve* Γ *bounds a minimal surface of the type of the* MÖBIUS *strip.* The geometrical interpretation is similar to that in the two-contour case.

VI.34. The preceding condition presupposes that $\overline{m}(\Gamma)$ and $m(\Gamma)$ are finite. In complete analogy with the two-contour case, DOUGLAS obtains the following generalization. Denote by $e(\Gamma)$ the maximum of $\overline{\lim}\,(m(\Gamma^{(n)}) - \overline{m}(\Gamma^{(n)}))$ for all sequences of JORDAN curves $\Gamma^{(n)} \to \Gamma$ such that $m(\Gamma^{(n)})$ and $\overline{m}(\Gamma^{(n)})$ are finite. *Whenever* $e(\Gamma) > 0$, Γ *bounds a minimal surface of the type of the* MÖBIUS *strip.*

VI.35. An application of the preceding criterion, considered by DOUGLAS, is not quite convincing. Suppose the given JORDAN curve Γ has upon the xy-plane a projection of the shape indicated in the figure. Denote by S the minimal surface of the type of the circular disc bounded by Γ. Then, according to DOUGLAS[1], S should evidently have the general form of the RIEMANN surface of $(x + iy)^{\frac{1}{2}}$, that is to say S should evidently have a branch-point, and from this fact DOUGLAS infers that Γ bounds a minimal surface of the type of the MÖBIUS strip. Well then, it is obvious that if the xy-projection of Γ is prescribed as in the figure, the xz-projection can be chosen as a simply covered star-shaped curve, and then (see III.10) *none* of the minimal surfaces of the type of the circular disc bounded by Γ will have a branch-point. It should also be observed that if a JORDAN curve Γ has an zy-projection as indicated in the figure, then Γ is certainly not knotted. Hence problem P_3 (see III.5) has a solution for Γ (see V.8). If a solution S of problem P_3 has a branch-point, then the order is at least 2 (see III.20). Every plane through a branch-point of S intersects therefore Γ in at least 6 distinct points (see III.8). Applying this remark to planes perpendicular to the xy-plane, it follows that whenever Γ has an xy-projection as indicated in the figure, then Γ bounds a minimal surface of the type of the circular disc which is free of branch-points.

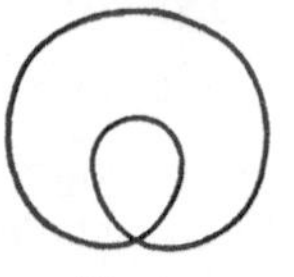

Fig. 1.

These remarks show the interest and the necessity of a thorough investigation of the singularities of minimal surfaces in their dependence upon the shape of the boundary curve.

(Closed Dec. 15, 1932.)

[1] One-sided minimal surfaces with a given boundary. Trans. Amer. Math. Soc. Vol. 34 (1932) pp. 733, 739, 753.

ERGEBNISSE DER MATHEMATIK
UND IHRER GRENZGEBIETE
HERAUSGEGEBEN VON DER SCHRIFTLEITUNG
DES
„ZENTRALBLATT FÜR MATHEMATIK"
FÜNFTER BAND
1

SUBHARMONIC FUNCTIONS

BY

TIBOR RADÓ

BERLIN
VERLAG VON JULIUS SPRINGER
1937

Preface.

A convex function f may be called *sublinear* in the following sense: if a linear function l is $\geqq f$ at the boundary points of an interval, then $l \geqq f$ in the interior of that interval also. If we replace the terms *interval* and *linear function* by the terms *domain* and *harmonic function*, we obtain a statement which expresses the characteristic property of *subharmonic functions* of two or more variables. This generalization, formulated and developed by F. RIESZ, immediately attracted the attention of many mathematicians, both on account of its intrinsic interest and on account of the wide range of its applications. If $f(z)$ is an analytic function of the complex variable $z = x + iy$, then $|f(z)|$ is subharmonic. The potential of a negative mass-distribution is subharmonic. In differential geometry, surfaces of negative curvature and minimal surfaces can be characterized in terms of subharmonic functions. The idea of a subharmonic function leads to significant applications and interpretations in the fields just referred to, and conversely, every one of these fields is an apparently inexhaustible source of new theorems on subharmonic functions, either by analogy or by direct implication. The purpose of this report is *first* to give a detailed account of those facts which seem to constitute the general theory of subharmonic functions, and *second* to present a selected group of facts which seem to be well adapted to illustrate the relationships between subharmonic functions and other theories. Roughly, Chapters I, II, III, V, VI are devoted to the first purpose, while Chapters IV and VII are devoted to the second one. The presentation is formulated for the case of two independent variables, but both the methods and the results remain valid in the general case, except for obvious modifications, unless the contrary is explicitly stated.

Subharmonic functions have a long and interesting history. F. RIESZ points out that various methods, due to POINCARÉ, PERRON, REMAK in potential theory and to HARTOGS and R. NEVANLINNA in the theory of functions of a complex variable, are based essentially on the idea of a subharmonic function. The reader should consult RIESZ [**4**], [**5**] for detailed historical references. Readers interested in the possibilities of applying subharmonic functions may read, for general information, RIESZ [**4**], [**5**], BECKENBACH-RADÓ [**1**], [**2**], EVANS [**4**], FROSTMAN [**1**].

As it has been observed above, potentials of negative mass-distributions are subharmonic functions, and essentially the converse is also true (see Chapter VI). Thus the theory of subharmonic functions may be interpreted as the study of such potentials based on a few *characteristic properties*, while the methods of potential theory are based on the *representation in terms of definite integrals*. It is very probable that the range of the theory of subharmonic functions, interpreted in this manner, will be considerably extended in the near future. For instance, *the sweeping-out process*, which is fundamental in the recent development of the theory of the capacity of sets (cf. EVANS [**4**], FROSTMAN [**1**]) could be easily interpreted in terms of harmonic majorants of subharmonic functions.

Historically, the first generalization of convex functions of a single variable is represented by the *convex functions of several variables*, characterized by the property of being *sublinear* on every straight segment within the domain of definition. While such functions are easily seen to be subharmonic, their theory was developed in connection with problems of an entirely different type. For this reason, the theory of these functions will be included among the topics discussed by W. FENCHEL in a subsequent report of this series.

The reviewer is indebted to G. C. EVANS and S. SAKS for valuable information which he had the privilege to use while preparing this report.

The Ohio State University, March 1937.

TIBOR RADÓ.

Contents.

Chapter I.

Definition and preliminary discussion of subharmonic functions.

1.1. Let $u(x,y)$ be a function in a domain G (connected open set), such that $-\infty \leqq u < +\infty$ in G. That is, $-\infty$ is an admissible value of u, while $+\infty$ is not. Such a function is *subharmonic* in G if it satisfies the following conditions (RIESZ [**5**], part I, p. 333).

a) u is not identically equal to $-\infty$ in G.

b) u is upper semi-continuous in G. That is, for every point (x_0, y_0) in G and for every number $\lambda > u(x_0, y_0)$ there exists a $\delta = \delta(x_0, y_0, \lambda) > 0$ such that $u(x,y) < \lambda$ for $[(x-x_0)^2 + (y-y_0)^2]^{1/2} < \delta$. Observe that for $u(x_0, y_0) = -\infty$ this condition implies that $u(x,y) \to -\infty$ for $(x,y) \to (x_0, y_0)$.

c) Let G' be any domain comprised in G together with its boundary B'. Let $H(x,y)$ be harmonic in G', continuous in $G' + B'$, and $H \geqq u$ on B'. Whenever these assumptions are satisfied, we also have $H \geqq u$ in G'.

Superharmonic functions are defined in a similar fashion. A function v is superharmonic in a domain G if the function $u = -v$ is subharmonic there. In the sequel we shall state the results only for subharmonic functions. In n-dimensional Euclidean space subharmonic functions are defined in exactly the same way as in the two-dimensional case. Clearly, a harmonic function is both subharmonic and superharmonic, and conversely.

1.2. For the sake of accuracy let us observe that F. RIESZ assumed that

a*) $u > -\infty$ on a set everywhere dense in G.

The apparently weaker condition a) in 1.1 was stated by EVANS [**4**], part I, p. 230. The following presentation, based partly on unpublished remarks of G. C. EVANS, will show that conditions a), b), c) are equivalent to conditions a*), b), c).

1.3. Condition b) will be used in the following way. Let S be a closed set comprised in G. Then condition b) implies (HAHN [**1**]) that there exists a sequence of functions φ_k with the following properties. α) φ_k is continuous on S. β) $\varphi_k \searrow u$ on S, where the symbol $\searrow$ indicates that $\varphi_1 \geqq \varphi_2 \geqq \cdots$. Conversely, the existence of such a sequence φ_k, for every choice of the closed set S in G, implies that u is upper semi-continuous in G. Take now a domain G' which is comprised in G together with its boundary B'. By what precedes, we have on B' a sequence of continuous functions φ_k such that $\varphi_k \searrow u$ on B'. Suppose that $G'+B'$ is a *Dirichlet region,* that is a region such that for every continuous function f on B' there exists a function H which is harmonic in G', continuous in $G'+B'$, and equal to f on B'. Denote by H_k the solution of the DIRICHLET problem for $G'+B'$ with the boundary condition $H_k=\varphi_k$ on B'. Then $\varphi_1 \geqq \varphi_2 \geqq \cdots$ implies that $H_1 \geqq H_2 \geqq \cdots$ in $G'+B'$. Since $H_k=\varphi_k \geqq u$ on B', it follows from condition c) in 1.1 that $H_k \geqq u$ in $G'+B'$. Summing up: for every Dirichlet region $G'+B'$ in G we have a sequence of functions H_k with the following properties. 1) H_k is continuous in $G'+B'$. 2) H_k is harmonic in G'. 3) $H_k \geqq u$ in $G'+B'$. 4. $H_k \searrow u$ on B'.

According to a fundamental theorem of HARNACK (KELLOGG [1], Chapter X), the property $H_1 \geqq H_2 \geqq \cdots$ implies that in G' the sequence H_k converges either to $-\infty$ everywhere or to a function h which is harmonic in G'. In the second case the convergence is uniform on every closed set in G'. The first case can be excluded as soon as $u>-\infty$ at a single point of G'. In the second case, $H_k \geqq u$ in $G'+B'$ implies that $\bar{h} \geqq u$ in G'.

Remark. If u is continuous, we can take $\varphi_k=u$, and h is then simply the solution of the Dirichlet problem for $G'+B'$ with the boundary condition $\bar{h}=u$ on B'. In the sequel, the reader interested only in continuous subharmonic functions should always consider this particular choice of φ_k. The reader interested in the general case should glance through the sections 5.1 to 5.4 at this time.

1.4. In the sequel we shall use the following theorems on integration quite frequently. Let there be given, on some range S (curve, domain, etc.) a sequence of functions F_n such that $F_n \searrow F$ on S and $\int_S F_n \geqq A$, where A is a finite constant independent of n (the integrals are taken in the sense of LEBESGUE). Then (see for instance SAKS [5], p. 63 and p. 83) the limit function F is also summable on S and we have $\int_S F=\lim \int_S F_n$.

Consider next a function $f(x,y)$ in a circular disc D: $(x-x_0)^2+(y-y_0)^2 \leqq r^2$. Introducing polar coordinates we have

$$\iint_D f(x,y)\,dx\,dy=\int_0^r\int_0^{2\pi} f(x_0+\varrho\cos\varphi,\, y_0+\varrho\sin\varphi)\,\varrho\,d\varrho\,d\varphi$$

$$=\int_0^r\left(\int_0^{2\pi} f(x_0+\varrho\cos\varphi,\, y_0+\varrho\sin\varphi)\,d\varphi\right)\varrho\,d\varrho,$$

where we know, by a theorem of TONELLI (SAKS [5], p. 75) that these formulas are certainly valid if $f(x, y)$ is measurable and $\geqq 0$ in D. More generally, these formulas are valid if f is bounded in one direction, say $f \leqq M$ on D, as it follows by applying the preceding remark to the function $M - f$ (quite exactly, whenever one of the three integrals involved exists, the other two exist also and the three integrals are equal to each other).

1.5. It will be convenient to use the following notations. $C(x_0, y_0; r)$, $D(x_0, y_0; r)$ will refer to the perimeter and to the interior respectively of the circle with centre (x_0, y_0) and radius r, while $R(x_0, y_0; r_1, r_2)$ will refer to the interior of the concentric ring bounded by the circles $C(x_0, y_0; r_1)$ and $C(x_0, y_0; r_2)$. If a function f, defined on $C(x_0, y_0; r)$, is summable as a function of the polar angle φ (where $x = x_0 + r\cos\varphi$, $y = y_0 + r\sin\varphi$), then we shall write

$$L(f;\, x_0, y_0;\, r) = \frac{1}{2\pi}\int_0^{2\pi} f(x_0 + r\cos\varphi,\, y_0 + r\sin\varphi)\, d\varphi\,.$$

Similarly, if f is defined and summable on $D(x_0, y_0; r)$, we shall write

$$A(f;\, x_0, y_0;\, r) = \frac{1}{r^2\pi}\iint_D f(x, y)\, dx\, dy\,.$$

We have the equivalent formula

$$A(f;\, x_0, y_0;\, r) = \frac{1}{r^2\pi}\iint_{\xi^2+\eta^2<r^2} f(x_0 + \xi,\, y_0 + \eta)\, d\xi\, d\eta\,.$$

$L(f;\, x_0, y_0;\, r)$ and $A(f;\, x_0, y_0;\, r)$ are the integral means of f on $C(x_0, y_0; r)$ and $D(x_0, y_0; r)$ respectively.

1.6. Throughout this Chapter u will denote a function which is subharmonic in a domain G. Suppose that the circle $C(x_0, y_0; r)$ is comprised in G together with its interior and also suppose that $u(x_0, y_0) > -\infty$. Then $L(u;\, x_0, y_0;\, r)$ exists and $u(x_0, y_0) \leqq L(u;\, x_0, y_0;\, r)$ (RIESZ [5], part I, p. 324). To see this, take, as in 1.3, a sequence H_k for the circular disc bounded by $C(x_0, y_0; r)$. We have then $u(x_0, y_0) \leqq H_k(x_0, y_0)$ and $H_k(x_0, y_0) = L(H_k;\, x_0, y_0;\, r)$ (see KELLOGG [1], p. 82). Hence $L(H_k;\, x_0, y_0;\, r) \geqq u(x_0, y_0) > -\infty$. By 1.4 it follows that $L(u;\, x_0, y_0;\, r)$ exists and that

$$u(x_0, y_0) \leqq \lim L(H_k;\, x_0, y_0;\, r) = L(u;\, x_0, y_0;\, r)\,.$$

1.7. Under the assumptions of 1.6 let us consider the disc $D(x_0, y_0; r)$ bounded by $C(x_0, y_0; r)$. We can then apply the result of 1.6 to $C(x_0, y_0; \varrho)$ for $0 < \varrho < r$, and we obtain by 1.4

$$A(u;\, x_0, y_0;\, r) = \frac{2}{r^2}\int_0^r L(u;\, x_0, y_0;\, \varrho)\, \varrho\, d\varrho \geqq \frac{2u(x_0, y_0)}{r^2}\int_0^r \varrho\, d\varrho = u(x_0, y_0)\,.$$

That is, if $u(x_0, y_0) > -\infty$, then (RIESZ [5], part II, p. 343) u is summable on every disc $D(x_0, y_0; r)$ completely interior to G and $u(x_0, y_0) \leqq A(u; x_0, y_0; r)$.

1.8. While condition a) in 1.1 states only that $u > -\infty$ for at least one point in G, it follows from 1.7 (according to an unpublished remark of G. C. EVANS) that $u > -\infty$ on a set which is everywhere dense in G. Indeed, if $u(x_0, y_0) > -\infty$, then the summability of u on the disc $D(x_0, y_0; r)$ implies that $u > -\infty$ on this disc with the possible exception of a set of two-dimensional LEBESGUE measure zero. Given then any other point (x', y') in G, we have clearly a finite number of discs $D(x_k, y_k; r_k)$, $k = 0, 1, \ldots, n$, completely interior to G, such that (x_{k+1}, y_{k+1}) is a point of $D(x_k, y_k; r_k)$ for which $u(x_{k+1}, y_{k+1}) > -\infty$, and such that $D(x_n, y_n; r_n)$ contains (x', y'). By 1.7 we have $u > -\infty$ almost everywhere on these discs and hence we have in the vicinity of (x', y') points (x^*, y^*) such that $u(x^*, y^*) > -\infty$.

1.9. u is summable on every disc $D(x_0, y_0; r)$ completely interior to G (RIESZ [5], part II, p. 343; cf. 1.2). Indeed, by 1.8 we have some disc $D(\bar{x}, \bar{y}; \bar{r})$ completely interior to G, such that $u(\bar{x}, y) > -\infty$ and such that $D(x_0, y_0; r)$ is comprised in $D(\bar{x}, \bar{y}; \bar{r})$. The assertion follows then immediately from 1.7.

1.10. u is summable on every measurable set S completely interior to G (by completely interior we mean that the limit points of S are also comprised in G). Indeed, by the HEINE-BOREL theorem we can cover the set $S + S'$, where S' is the set of the limit points of S, by a finite number of discs completely interior to G, and the assertion follows then from 1.9 (RIESZ [5], part II, p. 344).

1.11. u is summable, as a function of the polar angle, on every circle $C(x_0, y_0; r)$ comprised in G together with its interior (RIESZ [5], p. 334). This can be seen by the same reasoning as that used in 1.6, since the assumption $u(x_0, y_0) > -\infty$ was used there only to exclude the possibility $H_k \to -\infty$, and this is excluded now by 1.8.

1.12. u is summable, as a function of the polar angle, on every circle $C(x_0, y_0; r)$ comprised in G, even if the interior of $C(x_0, y_0; r)$ is not comprised in G (RIESZ [5], part I, p. 338). To see this, take a circle $C(x_0, y_0; r_1)$, $r_1 > r$, such that the ring $R(x_0, y_0; r, r_1)$ is comprised in G together with its boundary. As in 1.3, take a sequence H_k for this ring. By 1.4 the theorem is proved if we show that the sequence $L(H_k; x_0, y_0; r)$ is bounded from below.

Take any smooth JORDAN curve Γ in $R(x_0, y_0; r, r_1)$ which encloses $C(x_0, y_0; r)$. Then the integral

$$\int_{\Gamma} \frac{\partial H_k}{\partial n_e} ds,$$

where n_e refers to the outward normal of Γ, is independent of the choice of Γ (KELLOGG [1], p. 212). If we apply this to $C(x_0, y_0; \varrho)$, $r < \varrho < r_1$, then it follows that

$$\varrho \frac{d}{d\varrho} L(H_k; x_0, y_0; \varrho) = a_k$$

and hence

$$L(H_k; x_0, y_0; \varrho) = a_k \log\varrho + b_k$$

where a_k and b_k are constants. On account of the continuity of H_k in the closed ring this formula holds for $r \leqq \varrho \leqq r_1$. By 1.8 and 1.3, H_k converges in the open ring $R(x_0, y_0; r, r_1)$ to a harmonic function $\bar{h}$, the convergence being uniform, in particular, on every circle $C(x_0, y_0; \varrho)$, $r < \varrho < r_1$. Hence $L(H_k; x_0, y_0; \varrho) \to L(\bar{h}; x_0, y_0; \varrho)$ for $r < \varrho < r_1$. That is, the sequence $a_k \log\varrho + b_k$ has a finite limit for every ϱ such that $r < \varrho < r_1$. Clearly, this implies that a_k and b_k converge to finite limits a and b respectively. Then we have $L(H_k; x_0, y_0; r) \to a \log r + b$ and this implies that the sequence $L(H_k; x_0, y_0; r)$ is bounded from below.

1.13. Using the notations of 1.12, let us consider $L(u; x_0, y_0; \varrho)$ as a function of ϱ, $r \leqq \varrho \leqq r_1$. By the theorem of 1.12, $L(u; x_0, y_0; \varrho)$ actually exists. We observe that by 1.3 and 1.4 we have

$$L(u; x_0, y_0; r) = \lim L(H_k; x_0, y_0; r) = a \log r + b\,,$$

$$L(u; x_0, y_0; r_1) = \lim L(H_k; x_0, y_0; r_1) = a \log r_1 + b\,.$$

Consider now a third circle $C(x_0, y_0; \varrho)$, $r \leqq \varrho \leqq r_1$. Then $L(u; x_0, y_0; \varrho) \leqq L(H_k; x_0, y_0; \varrho)$ since $u \leqq H_k$ in the ring $R(x_0, y_0; r, r_1)$. Hence $L(u; x_0, y_0; \varrho) \leqq \lim L(H_k; x_0, y_0; \varrho) = a \log\varrho + b$. As $a \log\varrho + b$ is the (univocally determined) linear function of $\log\varrho$ which is equal to $L(u; x_0, y_0; \varrho)$ for $\varrho = r$ and $\varrho = r_1$, we have the following theorem (RIESZ [5], p. 338).

If the circular ring $0 \leqq \varrho_1^2 < (x - x_0)^2 + (y - y_0)^2 < \varrho_2^2$ is comprised in G, then $L(u; x_0, y_0; \varrho)$ is a convex function of $\log\varrho$ for $\varrho_1 < \varrho < \varrho_2$.

1.14. If we are willing to use somewhat more complicated tools than in the preceding sections, then we can obtain the following more comprehensive result (BRELOT [1], p. 14). If Γ is any sufficiently smooth JORDAN curve in G, then u is summable on Γ as a function of the arc-length (it is not necessary to assume that the interior of Γ is also comprised in G). We modify the proof of BRELOT slightly so as to obtain this theorem directly from the definition of a subharmonic function as given in 1.1. Let (x_0, y_0) be a point in G such that $u(x_0, y_0) > -\infty$, and assume that Γ does not pass through (x_0, y_0) [clearly, the case of curves passing through (x_0, y_0) can be settled then immediately]. We choose a second smooth JORDAN curve Γ_1 such that the doubly connected domain G' bounded by Γ and Γ_1 is comprised in G together with its boundary, and such that (x_0, y_0) is comprised in G'. As in 1.3, we take a sequence H_k for G'. Since $H_1 \geqq H_2 \geqq \cdots$ and since all these functions are continuous on $G' + \Gamma + \Gamma_1$, we have a finite constant M such that $u \leqq H_k \leqq M$, $k = 1, 2, \ldots$, in $G' + \Gamma + \Gamma_1$. If $\mathfrak{G}'(x, y)$ denotes GREEN's function for G' with pole at (x_0, y_0), we have, by applying a general formula (KELLOGG [1], p. 237) to the harmonic function $H_k - M$,

$$H_k(x_0, y_0) - M = \frac{1}{2\pi}\int_{\Gamma} (H_k - M)\frac{\partial \mathfrak{G}'}{\partial n_i} ds + \frac{1}{2\pi}\int_{\Gamma_1} (H_k - M]\frac{\partial \mathfrak{G}'}{\partial n_i} ds,$$

where n_i refers to the interior normal with respect to G'. We observe that $\partial \mathfrak{G}'/\partial n_i$ has a *positive* minimum $\mu > 0$ on $\Gamma + \Gamma_1$ (cf. BRELOT [1], p. 14). As $H_k - M \leqq 0$ on $\Gamma + \Gamma_1$, it follows that

$$H_k(x_0, y_0) - M \leqq \frac{\mu}{2\pi}\int_{\Gamma} (H_k - M)\, ds.$$

As $H_k(x_0, y_0) \geqq u(x_0, y_0)$, it follows finally that

$$\int_{\Gamma} H_k\, ds \geqq [u(x_0, y_0) - M]\frac{2\pi}{\mu} + M l,$$

where l is the length of Γ. That is, the sequence $\int_{\Gamma} H_k\, ds$ is bounded from below. By 1.3 and 1.4 it follows then that u is summable on Γ as function of the arc-length.

1.15. Let G' be a domain comprised in G together with its boundary B'. Suppose that H is harmonic in G', continuous in $G' + B'$ and $H \geqq u$ on B'. By condition c) in 1.1 we have then $H \geqq u$ in G' also. We shall see now that the sign of equality holds either everywhere or nowhere in G' (RIESZ [5], p. 331). Suppose there is some point (x_0, y_0) in G' such that $u(x_0, y_0) = H(x_0, y_0)$. If r is small, we have then, by 1.7,

$$H(x_0, y_0) = u(x_0, y_0) \leqq A(u; x_0, y_0; r) \leqq A(H; x_0, y_0; r) = H(x_0, y_0).$$

As $u \leqq H$, this clearly implies that $u \equiv H$ in the vicinity of (x_0, y_0). That is, the set of points in G' where $u = H$ is an open set. On account of the upper semi-continuity of u, the set of points in G' where $u < H$ is also open. As the first one of these sets is not empty by assumption, the second one must be empty (since the connected open set G' cannot be the sum of two non-overlapping open sets). Hence $u = H$ everywhere in G'. As u is upper semi-continuous and H is continuous and $\geqq u$ in $G' + B'$, it follows immediately that we have $u = H$ on the boundary of G' also.

1.16. As an immediate corollary of the preceding theorem we note the fact that u cannot have a local maximum at a point (x_0, y_0) in G, unless it reduces to a constant in the vicinity of (x_0, y_0), and that u cannot reach its absolute maximum in G unless it reduces to a constant in G.

Chapter II.

Integral means of subharmonic functions.

2.1. If u is subharmonic in a domain G, then $u(x_0, y_0) \leqq L(u; x_0, y_0; r)$, $u(x_0, y_0) \leqq A(u; x_0, y_0; r)$ for (x_0, y_0) in G and for sufficiently small r (see 1.6, 1.7, 1.9, 1.11). The question arises as to whether these relations are characteristic for subharmonic functions.

2.2. Denote by K the class of functions u which are defined in a given domain G and satisfy there the following conditions. α) $-\infty \leqq u < +\infty$ and $u \not\equiv -\infty$ in G. β) u is upper semi-continuous in G. Denote by K_1, K_2, K_3, K_4 the subclasses of K defined by the following additional requirements. A function u in K belongs to K_1 if for every point (x_0, y_0) in G with $u(x_0, y_0) > -\infty$ we have a $\varrho(x_0, y_0) > 0$ such that for $r < \varrho(x_0, y_0)$ the integral mean $L(u; x_0, y_0; r)$ exists and is $\geqq u(x_0, y_0)$. A function u in K belongs to K_2 if for every point (x_0, y_0) in G with $u(x_0, y_0) > -\infty$ there exists a sequence $r_n \to 0$, depending upon (x_0, y_0), such that $L(u; x_0, y_0; r_n)$ exists and is $\geqq u(x_0, y_0)$, $n = 1, 2, \ldots$. The classes K_3, K_4 are defined in the same way in terms of the integral mean $A(u; x_0, y_0; r)$.

2.3. On account of 1.6, 1.7, 1.9, 1.11 every function which is subharmonic in G belongs to all four classes K_1, K_2, K_3, K_4. Conversely (LITTLEWOOD [**1**], p. 189), a function u which belongs to any one of these classes is subharmonic in G. Since conditions a) and b) of 1.1 are satisfied by assumption, we have to verify only the following fact: if G' is a domain comprised in G together with its boundary B', and if H is continuous in $G' + B'$, harmonic in G', and $\geqq u$ on B', then $H \geqq u$ in G' also. If this were not so, then the function $u - H$, which is clearly upper semi-continuous in $G' + B'$, would reach a *positive* maximum M at an *interior* point (x_0, y_0) of $G' + B'$, and the set S of those points (x, y) in $G' + B'$ where $u - H = M$ would be a *closed* set *interior* to $G' + B'$. Since S and B' are closed sets, we have then on S a point (x_1, y_1) whose distance from B' would be a minimum. On every circle $C(x_1, y_1; r)$, with small r, we would have then a whole arc σ_r such that $u - H < M$ on σ_r. Since $u - H \leqq M$ on $C(x_1, y_1; r)$, it follows that

$$L(u; x_1, y_1; r) - H(x_1, y_1) = L(u - H; x_1, y_1; r) < M$$
$$= u(x_1, y_1) - H(x_1, y_1),$$

and hence $L(u; x_1, y_1; r) < u(x_1, y_1)$ for *all* small values of r for which $L(u; x_1, y_1; r)$ exists. A similar reasoning shows that $A(u; x_1, y_1; r) < u(x_1, y_1)$ for *all* small values of r for which $A(u; x_1, y_1; r)$ exists. These conclusions are in obvious contradiction to the assumption that u belongs to one of the classes K_1, K_2, K_3, K_4.

2.4. If u is subharmonic in the domain G, then for fixed (x_0, y_0) the integral mean $L(u; x_0, y_0; r)$ is an increasing function of r as long as the circle $C(x_0, y_0; r)$ is comprised in G together with its interior (RIESZ [5], part I, p. 338). To see this, suppose $C(x_0, y_0; r)$ satisfies the assumption of the theorem and take $r_1 < r$. As in 1.3, take a sequence H_k for the circular disc bounded by $C(x_0, y_0; r)$. We have then $L(H_k; x_0, y_0; r) = H_k(x_0, y_0) = L(H_k; x_0, y_0; r_1) \geqq L(u; x_0, y_0; r_1)$. By 1.3, 1.4 it follows for $k \to \infty$ that $L(u; x_0, y_0; r) \geqq L(u; x_0, y_0; r_1)$.

2.5. We already observed (see 1.13) that as long as the circle $C(x_0, y_0; r)$ remains in a circular ring $R(x_0, y_0; r_1, r_2)$ comprised in G, $L(u; x_0, y_0; r)$ is a convex function of $\log r$ for $r_1 < r < r_2$ and for fixed (x_0, y_0).

2.6. Under the assumptions of 2.5, $L(u; x_0, y_0; r)$ is a continuous function of r. Indeed, if $r_1 < r' < r'' < r_2$, then by reasons of upper semi-continuity u and therefore $L(u; x_0, y_0; r)$ is bounded from above for $r' \leqq r \leqq r''$, and a convex function which is bounded from above is continuous.

2.7. Suppose u is subharmonic in a disc $D(x_0, y_0; \bar{r})$. Then $L(u; x_0, y_0; r) \to u(x_0, y_0)$ for $r \to 0$ (RIESZ [5], part II, p. 344). Indeed, for $r \to 0$ the upper semi-continuity of u implies that $\overline{\lim}\, L(u; x_0, y_0; r) \leqq u(x_0, y_0)$, while by 1.6 we have $\underline{\lim}\, L(u; x_0, y_0; r) \geqq u(x_0, y_0)$ Clearly, the reasoning remains valid for $u(x_0, y_0) = -\infty$.

2.8. Similar theorems hold for

$$A(u; x_0, y_0; r) = \frac{2}{r^2} \int_0^r L(u; x_0, y_0; \varrho)\, \varrho\, d\varrho,$$

where it is assumed that the circle $C(x_0, y_0; r)$ is comprised togethei with its interior in a domain G where u is subharmonic. Since $L(u; x_0, y_0; \varrho)$ is a continuous function of ϱ for $0 \leqq \varrho \leqq r$ (see 2.6, 2.7), we can approximate the above integral by RIEMANN sums, and we obtain the relation (cf. MONTEL [2], p. 49)

$$A(u; x_0, y_0; r) = \lim_{n \to \infty} \sum_{k=1}^{n} \frac{2k}{n^2} L\left(u; x_0, y_0; \frac{k}{n} r\right).$$

As $L(u; x_0, y_0; \varrho)$ is an increasing function of ϱ, it follows immediately that $A(u; x_0, y_0; r) \leqq L(u; x_0, y_0; r)$ (cf. 3.25).

2.9. Under the assumptions of 2.8, $A(u; x_0, y_0; r)$ is an increasing function of r (RIESZ [5], part II, p. 344). This is obvious since the RIEMANN sums used in 2.8 are increasing functions of r by 2.4.

2.10. Under the assumptions of 2.8, we have $A(u; x_0, y_0; r) \to u(x_0, y_0)$ for $r \to 0$ (RIESZ [5], part II, p. 344). The proof is the same as in 2.7.

2.11. Under the assumptions of 2.8, $A(u; x_0, y_0; r)$ is a convex function of $\log r$ (essentially MONTEL [2], p. 49). This is obvious since the RIEMANN sums used in 2.8 are convex functions of $\log r$ by 2.5.

2.12. We shall say that a function u is of class PL in a domain G if $u \geqq 0$ and if $v = \log u$ is subharmonic there. It is understood that we put $v = -\infty$ for points where $u = 0$. If u is of class PL in G, then u is subharmonic in G (while the converse is obviously false). Indeed, the subharmonic character of $v = \log u$ clearly implies that u satisfies conditions a) and b) of 1.1. Take then any point (x_0, y_0) in G and a small r. As v is subharmonic by assumption, we have $v(x_0, y_0) \leqq L(v; x_0, y_0; r)$ and consequently

$$u(x_0, y_0) = e^{v(x_0, y_0)} \leqq e^{L(v; x_0, y_0; r)} \leqq L(u; x_0, y_0; r).$$

Hence u is subharmonic by 2.3. We used here the inequality

$$\frac{1}{2\pi}\int_0^{2\pi} \log f(\varphi)\, d\varphi \leqq \log \frac{1}{2\pi}\int_0^{2\pi} f(\varphi)\, d\varphi, \qquad f \geqq 0,$$

which is valid whenever the integrals involved have a meaning in the sense of LEBESGUE (for a very elegant proof, see RIESZ [7]).

2.13. If $u \not\equiv 0$ is $\geqq 0$ and upper semi-continuous in G, then u is of class PL there if and only if $u e^h$ is subharmonic for every choice of h in every subdomain G' in which h is harmonic (BECKENBACH [1], for continuous u; the following proof for general u is based on unpublished remarks of S. SAKS). The necessity of the condition being obvious by 2.12, let us prove that the condition is sufficient. Let G' be any domain comprised in G together with its boundary B'. Let H be continuous in $G' + B'$, harmonic in G', and $v = \log u \leqq H$ on B'. By assumption $u e^{-H}$ is subharmonic in G', and $u e^{-H}$ is upper semi-continuous even in $G' + B'$, since u is upper semi-continuous and $e^{-H} > 0$ is continuous there. The reasoning of 1.15 applies therefore to $u e^{-H}$ and since $u e^{-H} \leqq 1$ on B', we obtain $u e^{-H} \leqq 1$ in G' and finally $v = \log u \leqq H$ in G'. That is, $v = \log u$ is subharmonic in G, since v clearly satisfies conditions a) and b) in 1.1 also.

2.14. If u_1, u_2 are subharmonic in G, then $u_1 + u_2$ is clearly also subharmonic in G, while $u_1 u_2$ will generally not be subharmonic there. On the other hand, the class PL is closed both under addition and multiplication (PRIVALOFF [4]; the following proof is due to S. SAKS). That is, if u_1, u_2 are of class PL in G, then $v = u_1 u_2$ and $w = u_1 + u_2$ are also of class PL. For v this is obvious. As to w, consider any function h which is harmonic in a subdomain G' of G. By 2.13, $u_1 e^h$ and $u_2 e^h$ are subharmonic in G'. Hence $u_1 e^h + u_2 e^h = w e^h$ is also subharmonic in G'. By 2.13 it follows that $\log w$ is subharmonic in G.

2.15. For fixed (x_0, y_0) the function

$$\log r = \log r(x, y; x_0, y_0) = \begin{cases} \log[(x-x_0)^2 + (y-y_0)^2]^{1/2} & \text{for } (x, y) \neq (x_0, y_0), \\ -\infty & \text{for } (x, y) = (x_0, y_0) \end{cases}$$

is a subharmonic function of (x, y) in the whole plane. This follows from the fact that $\log r$ is harmonic for $(x, y) \neq (x_0, y_0)$, while at (x_0, y_0) both the value and the limit of $\log r$ are equal to $-\infty$. If $\alpha > 0$ is a constant, then $\alpha \log r$ is clearly also subharmonic. That is, r^α is of class PL in the whole plane, for $\alpha > 0$. In the case of three independent variables, for instance, we would have

$$-\frac{1}{r} = -\frac{1}{r(x, y, z; x_0, y_0, z_0)}$$

$$= \begin{cases} -\dfrac{1}{[(x-x_0)^2 + (y-y_0)^2 + (z-z_0)^2]^{1/2}} & \text{for } (x, y, z) \neq (x_0, y_0, z_0), \\ -\infty \quad \text{for} \quad (x, y, z) = (x_0, y_0, z_0) \end{cases}$$

as the simplest unbounded subharmonic function.

2.16. If u is of class PL in a ring $R(x_0, y_0; \varrho_1, \varrho_2)$, then $\log L(u; x_0, y_0; \varrho)$ is a convex function of $\log \varrho$ for $\varrho_1 < \varrho < \varrho_2$ (RIESZ [5], part I, p. 339). To show that a function $f(\varrho)$ has the property that $\log f(\varrho)$ is a convex function of $\log \varrho$, it is sufficient to show that $\varrho^\alpha f(\varrho)$ is a convex function of $\log \varrho$ for every $\alpha > 0$ (see for instance RIESZ [1], p. 6). Let us take any $\alpha > 0$. Put $r = [(x-x_0)^2 + (y-y_0)^2]^{1/2}$. Then, by 2.15 and 2.14, $r^\alpha u$ is of class PL in the ring and hence, by 2.12, $r^\alpha u$ is subharmonic in the ring. By 2.5, $L(u r^\alpha; x_0, y_0; \varrho) = \varrho^\alpha L(u; x_0, y_0; \varrho)$ is therefore a convex function of $\log \varrho$ for $\varrho_1 < \varrho < \varrho_2$.

2.17. If u is of class PL in a disc $D(x_0, y_0; \bar{r})$, then $\log A(u; x_0, y_0; r)$ is a convex function of $\log r$ for $0 < r < \bar{r}$ (essentially MONTEL [2], p. 48). This is obvious since the RIEMANN sums used in 2.8 have (by 2.16) the property that their logarithms are convex functions of $\log r$ for $0 < r < \bar{r}$.

2.18. For various purposes it is important to approximate general subharmonic functions by smooth subharmonic functions. We shall use the following terminology. A function $f(x, y)$ is of class $K^{(0)}$ in a domain G if it is continuous there, and it is of class $K^{(n)}$, $n \geqq 1$, if its derivatives of the first n orders are also continuous.

2.19. Let u be subharmonic in a domain G. Consider a domain G' contained in G together with its boundary B'. Put

$$A_r(x, y; u) = A(u; x, y; r) = \frac{1}{r^2 \pi} \iint_{\xi^2 + \eta^2 < r^2} u(x + \xi, y + \eta) \, d\xi \, d\eta$$

(see RIESZ [5], part II, p. 343 and p. 345 for historical references concerning the use of these approximating functions in the theory of harmonic and subharmonic functions). For r fixed and sufficiently small, $A_r(x, y; u)$ is a function of (x, y) which is defined and continuous in G' (cf. 1.10). By 2.1 we have $u(x, y) \leqq A_r(x, y; u)$ in G'. As u is bounded from above on every closed set in G, the theorem of TONELLI, referred to in 1.4, may be used to justify the changes in the order of integrations which we are going to carry out. First, from $u(x,y) \leqq A_r(x,y;u)$ we obtain by integration $A_{r_1}(x,y;u) \leqq A_{r_1}(x,y;A_r) = A_r(x,y;A_{r_1})$. That is (see 2.3) $A_{r_1}(x,y;u)$ is subharmonic. If we put $A_{r_1,r_2}(x,y;u) = A_{r_2}(x,y;A_{r_1})$ and so on, then it follows generally that for $n \geqq 1$ the function $A_{r_1,r_2,\ldots r_n}(x,y;u)$ is continuous and subharmonic in G' for small values of $r_1, r_2, \ldots, r_n$ and that $u(x,y) \leqq A_{r_1}(x,y;u) \leqq A_{r_1,r_2}(x,y;u) \leqq \ldots$. In particular, if we put $A_r^{(n)}(x,y;u) = A_{r,r,\ldots,r}(x,y;u)$, then for small r the function $A_r^{(n)}(x,y;u)$ is continuous and subharmonic in G and we have there $u(x,y) \leqq A_r^{(n)}(x,y;u)$.

By 2.8 we have $A_r(x,y;u) \leqq A_s(x,y;u)$ for $r \leqq s$. By repeated integration we obtain generally $A_r^{(n)}(x,y;u) \leqq A_s^{(n)}(x,y;u)$ for $r \leqq s$. Finally, by the same reasoning as in 2.7, we obtain $A_r^{(n)}(x,y;u) \to u(x,y)$ for $r \to 0$.

2.20. If $f(x,y)$ is *continuous* in a domain G, then for fixed r the function $A_r(x,y;f) = A(f;x,y;r)$ is easily seen to have continuous derivatives of the first order in the portion of G where it is defined. Indeed, the *four-step rule* for differentiation leads immediately to the formulas

$$\frac{\partial A_r(x,y;f)}{\partial x} = \frac{1}{r\pi}\int_0^{2\pi} f(x + r\cos\varphi,\ y + r\sin\varphi)\cos\varphi\, d\varphi,$$

$$\frac{\partial A_r(x,y;f)}{\partial y} = \frac{1}{r\pi}\int_0^{2\pi} f(x + r\cos\varphi,\ y + r\sin\varphi)\sin\varphi\, d\varphi,$$

which show the continuity of the first derivatives. If f itself has continuous derivatives $f_x = p$, $f_y = q$ of the first order in G, then we have simply

$$\frac{\partial A_r(x,y;f)}{\partial x} = A_r(x,y;p), \qquad \frac{\partial A_r(x,y;f)}{\partial y} = A_r(x,y;q),$$

and the preceding argument shows that $A_r(x,y;f)$ has continuous derivatives of the second order. Generally, if f is of class $K^{(n)}$, then $A_r(x,y;f)$ is of class $K^{(n+1)}$. Applying this to the function $A_r^{(n)}(x,y;u)$ of 2.19, it follows that $A_r^{(n)}(x,y;u)$ is of class $K^{(n-1)}$ in the portion of G in which it is defined.

2.21. For easier reference we sum up the preceding remarks in the following *approximation theorem*. If u is subharmonic in a domain G, then the sequence $u_k^{(3)}(x,y) = A_{\frac{1}{k}}^{(3)}(x,y;u)$, $k = 1, 2, \ldots$, has the following properties. Let G' be any domain comprised in G together with its boundary. Then for large k the function $u_k^{(3)}$ is defined, subharmonic and of class $K^{(2)}$ in G' and $u_k^{(3)} \searrow u$ in G'.

Actually, the integral means $A_r^{(n)}(x,y;u)$ are smoother than it appears from the preceding statements. More precise information could be obtained easily from 6.22.

2.22. For continuous u it follows from 2.19 and 2.20 that for large k the function $u_k^{(2)}(x,y) = A_{1/k}^{(2)}(x,y;u)$ is already of class $K^{(2)}$ in G' and that $u_k^{(2)} \to u$ *uniformly* in G'.

2.23. Suppose that the function u of 2.21 happens to be *harmonic* in a domain G' comprised in G together with its boundary B'. Take any closed set S in G' and denote by δ the shortest distance of S and of B'. Then for $r < 1/(3\delta)$ we have, by the mean-value property of harmonic functions, $u_k^{(3)} = u$ on S and hence $\Delta u_k^{(3)} = 0$ on S, $\Delta = \partial^2/\partial x^2 + \partial^2/\partial y^2$.

2.24. Denote by S a closed bounded set in the domain G in which u is subharmonic. Then the functions $u_k^{(3)}$ of 2.21 satisfy for large k an inequality

$$0 \leqq \iint_S \Delta u_k^{(3)}(x,y)\,dx\,dy < M,$$

where M is a finite constant (RIESZ [5], part II, p. 353). To see this, observe that S can be covered by a finite number of closed circular discs (HEINE-BOREL theorem) and that therefore it is sufficient to consider the case when S is a closed circular disc D, with radius r and centre (x_0, y_0), comprised in G. We have then by GREEN's identity

$$\iint_D \Delta u_k^{(3)}(x,y)\,dx\,dy = \int_{C_r} \frac{\partial u_k^{(3)}}{\partial u_e}\,ds = \frac{dL(u_k^{(3)}; x_0, y_0; r)}{d\log r}, \quad C_r = C(x_0, y_0; r).$$

Denote by I_k the common value of these expressions. Take r_1 slightly larger than r. Write $L_k(r)$ for $L(u_k^{(3)}; x_0, y_0; r)$ and $L(r)$ for $L(u; x_0, y_0; r)$. Since $L_k(r)$ is a convex function of $\log r$ by 2.5 and 2.21, we have

$$I_k \leqq \frac{L_k(r_1) - L_k(r)}{\log r_1 - \log r}.$$

By 2.21, 1.4, 2.4 it follows for $k \to \infty$ that

$$0 \leqq \overline{\lim}\, I_k \leqq \frac{L(r_1) - L(r)}{\log r_1 - \log r} < +\infty$$

and the theorem is proved.

Chapter III.

Criteria and constructions for subharmonic functions.

3.1. If u is of class $K^{(2)}$ (see 2.18) in a domain G and if $\Delta u = \partial^2 u/\partial x^2 + \partial^2 u/\partial y^2 > 0$, then u is subharmonic. Similarly, if $\Delta u < 0$, then u is superharmonic. Suppose, for instance, that $\Delta u > 0$ in G. Let G' be a domain comprised in G together with its boundary B'. Suppose that H is continuous in $G' + B'$, harmonic in G', and $\geqq u$ on B'. We have to show that $H \geqq u$ in G' also. Suppose this is not true. Then $v = u - H$ reaches its maximum at an interior point (x_0, y_0), and we should have there $\Delta v \leqq 0$, while by assumption $\Delta v = \Delta u - \Delta H = \Delta u > 0$ in G.

3.2. If u is of class $K^{(2)}$ in G, then u is subharmonic in G if and only if $\Delta u \geqq 0$ there (RIESZ [**5**], part I, p. 335). Proof. Suppose first that u is subharmonic in G. If $\Delta u < 0$ at some point (x_0, y_0) of G, then u is superharmonic in a vicinity G' of (x_0, y_0), on account of 3.1. Then u is both subharmonic and superharmonic in G', and hence (see 1.1) u is harmonic there. This is impossible, since $\Delta u < 0$ in the vicinity of (x_0, y_0). Suppose second that $\Delta u \geqq 0$ in G. If ε is a positive constant, then the function $u^* = u + \varepsilon(x^2 + y^2)$ satisfies the condition $\Delta u^* > 0$ in G. Hence u^* is subharmonic in G by 3.1. For $\varepsilon \to 0$ the function u^* converges to u uniformly in G, and the subharmonic character of u is then a consequence of the following theorem.

3.3. Let u be defined in a domain G. Suppose that there exists a sequence u_k with the following properties. If G' is any domain comprised in G together with its boundary, then for large k the function u_k is defined and subharmonic in G' and $u_k \to u$ uniformly in G'. Then u is subharmonic in G. Briefly, the uniform limit of subharmonic functions is subharmonic (RIESZ [**5**], part I, p. 335). Observe first that the assumptions clearly imply the upper semi-continuity of u. If (x_0, y_0) is any point in G, then we have, by 2.1, for large k and small r, $u_k(x_0, y_0) \leqq L(u_k; x_0, y_0; r)$ and for $k \to \infty$ it follows that $u(x_0, y_0) \leqq L(u; x_0, y_0; r)$. The subharmonic character of u follows then from 2.3.

3.4. If a, b are two real numbers, including $\pm\infty$, then $\overline{a, b}$ will denote the larger one of a, b if $a \neq b$ and the common value of a, b if $a = b$. We have then the theorem: if u_1, u_2 are subharmonic in G, then their upper envelope $u = \overline{u_1, u_2}$ is also subharmonic in G (RIESZ [**5**], part I, p. 335). Proof. u clearly satisfies conditions a) and b) in 1.1. Let $C(x_0, y_0; r)$ be any circle comprised in G together with its interior. As u_1 and u_2 are upper semi-continuous, we have a finite constant M such that $u_1 < M$, $u_2 < M$ on $C(x_0, y_0; r)$. We have then also $u < M$ on $C(x_0, y_0; r)$, while by definition $u \geqq u_1$. Thus u is comprised between two functions, namely u_1 and M, which are summable on $C(x_0, y_0; r)$ as functions of the polar angle. Hence u is also summable on $C(x_0, y_0; r)$ as a function of the polar angle. At (x_0, y_0) we have either $u = u_1$ or $u = u_2$. If $u(x_0, y_0) = u_1(x_0, y_0)$, for instance, then $u(x_0, y_0) = u_1(x_0, y_0) \leqq L(u_1; x_0, y_0; r) \leqq L(u; x_0, y_0; r)$. Hence u is subharmonic by 2.3. A similar reasoning shows that the upper envelope of any finite number of subharmonic functions is subharmonic.

3.5. Let F denote a family of infinitely many functions which are subharmonic in a domain G. Suppose that F is a normal family (that is, every infinite sequence of functions of F contains a uniformly convergent subsequence). If $\bar{u}(x, y)$ denotes the largest cluster-value of the values at (x, y) of all the functions of F, then $\bar{u}(x, y)$ is subharmonic in G (MONTEL [**2**], p. 38, for continuous functions; MALCHAIR [**1**], p. 11, for general subharmonic functions). The proof is similar to that in 3.4.

3.6. Let there be given, in a domain G, a sequence u_n with the following properties. If G' is any domain comprised in G together with its boundary, then for large n the functions $u_n, u_{n+1}, \dots$ are defined and subharmonic in G' and $u_n \geqq u_{n+1} \geqq \cdots$ in G'. Then either $u_n \to -\infty$ in G or u_n converges in G to a subharmonic function (RIESZ [**5**], part I, p. 335; cf. 1.2). Proof. Put $\lim u_n = u$ and suppose that $u \not\equiv -\infty$ in G. Clearly, $-\infty \leqq u < +\infty$ and u is upper semi-continuous in G. Let (x_0, y_0) be any point in G such that $u(x_0, y_0) > -\infty$. We have then, for large n and small r, $-\infty < u(x_0, y_0) \leqq u_n(x_0, y_0) \leqq L(u_n; x_0, y_0; r)$ and this implies, by 1.4, that $L(u; x_0, y_0; r)$ exists and is $\geqq u(x_0, y_0)$. Hence u is subharmonic by 2.3.

3.7. Suppose that u is upper semi-continuous and $-\infty \leqq u < +\infty$ in a domain G and that for every (x_0, y_0) in G the integral mean $L(u; x_0, y_0; r)$ exists for small r. If

$$\overline{\lim_{r \to 0}} \frac{1}{r^2} (L(u; x_0, y_0; r) - u(x_0, y_0)) \geqq 0$$

for every point (x_0, y_0) in G, then u is subharmonic in G (SAKS [**3**], p. 190; this is a generalization of a theorem of BLASCHKE [**1**] on harmonic functions). Proof. Observe that the function $u_n = u + x^2/n$, n a positive integer, belongs to the class K_2 defined in 2.2 and apply 2.3 and 3.6. A similar reasoning shows that if

$$\overline{\lim_{r \to 0}} \frac{1}{r^2} (A(u; x_0, y_0; r) - u(x_0, y_0)) \geqq 0$$

for every point (x_0, y_0) in G, then u is subharmonic in G (SZPILRAJN [**1**], p. 589). For further theorems of this type see the remarks of SAKS ([**4**], p. 382), and see also KOZAKIEWICZ ([**1**], pp. 5—6).

3.8. Combining 2.21, 3.6 and 3.2 we see that the class of subharmonic functions consists *first* of all functions with continuous second derivatives which satisfy the condition $\Delta = \partial^2/\partial x^2 + \partial^2/\partial y^2 \geqq 0$, and *second* of the limits of decreasing sequences of such functions (limits $\equiv -\infty$ being excluded).

3.9. Similarly, by 2.22, the class of *continuous* subharmonic functions consists *first* of the functions with $\Delta \geqq 0$ as before, and *second* of the *uniform* limits of such functions.

3.10. If $u_1, u_2, \ldots, u_n$ are subharmonic in G and if $\alpha_1, \alpha_2, \ldots, \alpha_n$ are non-negative constants, then obviously $\alpha_1 u_1 + \alpha_2 u_2 + \cdots + \alpha_n u_n$ is also subharmonic in G (RIESZ [**5**], part I, p. 335).

3.11. The next few theorems will be concerned with relations between subharmonic functions and convex functions. These theorems have the common feature that they can be proved by very simple computations if the functions involved are sufficiently smooth. It is then natural to treat the general case by approximation in terms of integral means (see 2.21). As a matter of fact, in the theory of subharmonic functions this method of approximation was first used in connection with a problem of this type (the theorem discussed in 3.12).

3.12. Let $u \geqq 0$ be upper semi-continuous and $< +\infty$ in a domain G. Then $v = \log u$ is subharmonic in G if and only if $e^{\alpha x + \beta y} u$ is subharmonic there for every choice of the constants α, β (MONTEL [**2**], p. 39 for smooth u; RADÓ [**2**], for continuous u), Proof. The necessity of the condition follows immediately from 2.12. To prove the sufficiency, suppose that $e^{\alpha x + \beta y} u = w$ is subharmonic for every choice of the constants α, β and call this property *the property* (M). If u is positive and of class $K^{(2)}$ in G, then we have $\Delta w \geqq 0$. Explicitly:

$$\Delta w = e^{\alpha x + \beta y} [\Delta u + 2\alpha u_x + 2\beta u_y + (\alpha^2 + \beta^2) u] \geqq 0$$

for every choice of α, β. As the quantity in the bracket is a quadratic function of α, β, we obtain readily the inequality $u \Delta u - (u_x^2 + u_y^2) \geqq 0$, which shows that $\log u$ is subharmonic. Indeed, we have $\Delta \log u = [u \Delta u - (u_x^2 + u_y^2)]/u^2$. If u, still of class $K^{(2)}$, satisfies only the condition $u \geqq 0$, consider first $u_n = u + 1/n$, n a positive integer, and apply 3.3. Suppose now that u has only the properties specified in the statement of the theorem. It follows then immediately that the function $A_r^{(3)}(x, y; u)$ of 2.19 also possesses the property (M). As $A_r^{(3)}(x, y; u)$ is of class $K^{(2)}$, $\log A_k^{(3)}(x, y; u)$ is subharmonic by what precedes, and as $\log A_r^{(3)}(x, y; u) \searrow \log u$ for $r \searrow 0$, the subharmonic character of $\log u$ follows by 3.6.

3.13. Suppose that $f(t)$ is convex and increasing (and therefore continuous) for $t_1 < t < t_2$ and that $u(x, y)$ is subharmonic in a domain G. If $t_1 < u < t_2$ in G, then $v = f(u)$ is subharmonic in G (MONTEL [**2**], p. 42, for smooth functions; BRELOT [**1**], p. 16, for the general case). Proof. If f and u are smooth, we have $\Delta v = f''(u)(u_x^2 + u_y^2) + f'(u) \Delta u \geqq 0$, since $f' \geqq 0$, $f'' \geqq 0$, $\Delta u \geqq 0$ by assumption. The general case can be treated either by approximation (BRELOT, l. c.) or also directly as follows. Let (x_0, y_0) be any point in G. Since u is subharmonic, we have $u(x_0, y_0) \leqq L(u; x_0, y_0; r)$ for small r. Since f is increasing, it follows that $v(x_0, y_0) = f(u(x_0, y_0)) \leqq f(L(u; x_0, y_0; r))$. Since f is convex, we have $f(L(u; x_0, y_0; r)) \leqq L(f(u); x_0, y_0; r) = L(v; x_0, y_0; r)$ (see for instance PÓLYA-SZEGŐ [**1**], p. 52, problem 71, where the inequality is stated in a somewhat less general form). Thus $v(x_0, y_0) \leqq L(v; x_0, y_0; r)$ and hence v is subharmonic by 2.3 (the further assumptions stated there being obviously satisfied in the present case).

3.14. Suppose that $f(t)$ is convex and continuous for $t_1 < t < t_2$ and that $h(x, y)$ is harmonic in a domain G. If $t_1 < h < t_2$ in G, then $v = f(h)$ is subharmonic in G (see 3.13 for references). The proof is the same as in 3.13.

3.15. The theorem of 3.12 can be restated as follows. Suppose that v is upper semi-continuous in a domain G and satisfies there the conditions $-\infty \leqq v < +\infty$, $v \not\equiv -\infty$. If $e^{\alpha x + \beta y + v}$ is subharmonic for every choice of the constants α, β, then v is subharmonic, and the converse is also true. Kierst (see Saks [3], p. 187) raised the following question. For what functions $f(t)$ is it true that whenever $f(\alpha x + \beta y + v)$ is subharmonic for every choice of the constants α, β, then it follows that v is subharmonic. According to 3.12, $f(t) = e^t$ is such a function. Kierst found the following curious theorem. If $f(t)$ for $-\infty < t < +\infty$ and $v(x, y)$ for (x, y) in a domain G have continuous derivatives of the second order, if $f'(t) > 0$, and if $f(\alpha x + \beta y + v)$ is subharmonic in G for every choice of the constants α, β, then v is subharmonic in G. That is, if we restrict ourselves to smooth functions, then every strictly increasing function $f(t)$ has the desired property. Proof. Put $w = f(\alpha x + \beta y + v)$. The assumption that w is subharmonic is expressed by the inequality

$$\Delta w = f''(\alpha x + \beta y + v)\,[(v_x + \alpha)^2 + (v_y + \beta)^2] + f'(\alpha x + \beta y + v)\,\Delta v \geqq 0.$$

Consider any point (x_0, y_0) in G and choose $\alpha = -v_x(x_0, y_0)$, $\beta = -v_y(x_0, y_0)$. As $f' > 0$, it follows that $\Delta v \geqq 0$ at (x_0, y_0). Hence v is subharmonic by 3.2. It is not known at present whether the assumptions concerning the smoothness of f and v are necessary for the validity of the theorem.

3.16. Saks [3] observed that the functions $f(t)$ for which the theorem of 3.15 is non-vacuous are less general than it would appear from the statement of that theorem. Suppose that 1) $f(t)$ is continuous with its first and second derivatives for $-\infty < t < +\infty$, 2) $f'(t) > 0$ and 3) there exists, in some domain G, a function $v(x, y)$ of class $K^{(2)}$ such that $f(\alpha x + \beta y + v)$ is subharmonic for every choice of the constants α, β. Then $f(t)$ is convex. As in 3.15, the proof follows by a simple discussion of the explicit expression of $\Delta f(\alpha x + \beta y + v)$. Again, it is not known whether the theorem remains true without the restrictions concerning the smoothness of f and v. Saks (l. c.) modified the problem by introducing a third parameter γ, and studied the situation where $f(\alpha x + \beta y + \gamma + v)$ is subharmonic for every choice of the constants α, β, γ. He found that as a consequence of the increased number of the parameters the assumptions concerning the derivatives of f and v can be dropped. He obtained the following theorem. Suppose that 1) $f(t)$ is continuous for $-\infty < t < +\infty$, 2) $v(x, y)$ is continuous in a domain G, and 3) $f(\alpha x + \beta y + \gamma + v)$ is subharmonic in G for every choice of the constants α, β, γ. Then $f(t)$ is necessarily convex and furthermore one of the following four statements is true. I. $f(t)$ is constant. II. $v(x, y)$ is harmonic. III. $v(x, y)$ is subharmonic in G and $f(t)$ is increasing for $-\infty < f < +\infty$ [that is, $f(t_1) \leqq f(t_2)$ for $t_1 < t_2$]. IV. $v(x, y)$ is superharmonic in G and $f(t)$ is decreasing for $-\infty < t < +\infty$. If f and v are sufficiently smooth, then this theorem can be proved by a simple argument similar to that used in 3.15. The hope that the general case can be treated by approximations has not materialized so far. At any rate, the proof given by SAKS is essentially a direct proof. The ingenious details of the proof cannot be reproduced here.

3.17. Suppose that 1) $f(t)$ is continuous for $t_1 < t < t_2$, 2) $h(x, y)$ is harmonic in a domain G, 3) $t_1 < h < t_2$ in G, and 4) $u = f(h)$ is subharmonic in G. Then $f(t)$ is convex in the interval $m < t < M$, where m and M denote the greatest lower bound and the least upper bound of h in G (MONTEL [2], p. 43, under certain restrictions; SAKS [**2**], for the general case). Proof. It is clearly assumed that h is not constant. If the theorem is false, then there exists a linear function $at + b$ such that $g(t) = f(t) + at + b$ reaches a proper local maximum at a certain point t_0, $m < t_0 < M$. That is, $g(t) \leqq g(t_0)$ and $g(t)$ is not constant in the vicinity of t_0. The function h takes on the value t_0 at some point (x_0, y_0) in G, since $m < t_0 < M$, and h takes on all values t close to t_0 in the vicinity of (x_0, y_0), since h is not constant. The function $u = g(h) = f(h) + ah + b$ has then a local maximum at (x_0, y_0) without reducing to a constant in the vicinity of (x_0, y_0). On account of 1.16 this is however impossible since u is clearly subharmonic.

3.18. We proceed to quote a few applications of the preceding theorems. Let u be subharmonic in a circular disc $D(x_0, y_0; \varrho)$. Put $[(x - x_0)^2 + (y - y_0)^2]^{1/2} = r$ and consider the function $l(x, y) = L(u; x_0, y_0; r)$. Then $l(x, y)$ is constant on every circle $C(x_0, y_0; r)$, $0 < r < \varrho$, and we can write $l(x, y) = \lambda(\log r)$. By 2.6, $\lambda(\log r)$ is a continuous, increasing and convex function of $\log r$ for $0 < r < \varrho$. Hence, by 3.13, $l(x, y)$ is subharmonic in $D(x_0, y_0; \varrho)$, except possibly at (x_0, y_0). To discuss the point (x_0, y_0), observe that $l(x, y)$ is continuous there by 2.6 and 2.7, and that $L(l; x_0, y_0; \sigma) = L(u; x_0, y_0; \sigma) \geqq u(x_0, y_0) = l(x_0, y_0)$ since u is subharmonic. Hence $l(x, y)$ is subharmonic in $D(x_0, y_0; \varrho)$ (essentially MONTEL [2], p. 48).

3.19. Under the assumptions of 3.18 the function $a(x, y) = A(u; x_0, y_0; r)$ is also subharmonic in $D(x_0, y_0; \varrho)$ (see reference in 3.18). The proof is the same as in 3.18.

3.20. Under the assumptions of 3.18 the function u will reach a maximum $M(r)$ on every circle $C(x_0, y_0; r)$, $0 < r < \varrho$. Put $M(0) = u(x_0, y_0)$ and consider the function $\mu(x, y) = M(r)$, $r = [(x - x_0)^2 + (y - y_0)^2]^{1/2}$. Then $\mu(x, y)$ is subharmonic in $D(x_0, y_0; \varrho)$ (essentially MONTEL [**2**], p. 41). On account of 3.14 this theorem is, except for minor details, a consequence of the fact that $M(r)$ is a convex function of $\log r$ (essentially RIESZ [**1**]). To prove this property of $M(r)$ take r_1, r_2 such that $0 < r_1 < r_2 < \varrho$. We have to show that $M(r) \leqq a \log r + b$ for $r_1 < r < r_2$, where $a \log r + b$ is the linear function of $\log r$ which takes on the values $M(r_1)$, $M(r_2)$ for $r = r_1$, $r = r_2$ respectively. Consider the function $H(x, y) = a \log[(x - x_0)^2 + (y - y_0)^2]^{1/2} + b$. Then H is harmonic and we have $H \geqq u$ on the boundary of the circular ring $R(x_0, y_0; r_1 . r_2)$ by the definition of $M(r)$ and of a, b. Hence (see 1.1, condition c)) we have also $H \geqq u$ on $C(x_0, y_0; r)$ for $r_1 < r < r_2$ and consequently $M(r) \leqq a \log r + b$ for $r_1 < r < r_2$.

3.21. If in the statements of the theorems in 3.18 to 3.20 we assume that u is of class PL (see 2.12), then it follows immediately that the functions $l(x, y)$, $a(x, y)$, $\mu(x, y)$ are also of class PL (see 3.18 to 3.20 for references).

3.22. Suppose that 1) $u \not\equiv -\infty$, 2) $-\infty \leqq u < +\infty$ and 3) u is upper semi-continuous in a domain G. If u is a convex function of x for every fixed value of y and a convex function of y for every fixed value of x, then u is subharmonic in G (MONTEL [**2**], p. 37, for continuous u; MALCHAIR [**1**], p. 7, for the general case). Proof. Take any circle $C(x_0, y_0; r)$ comprised in G together with its interior. Take a sequence of functions g_ν such that g_ν is continuous and $g_\nu \searrow u$ for $(x - x_0)^2 + (y - y_0)^2 = r^2$ (cf. 1.3). The convexity properties of u and the inequality $u \leqq g_\nu$ imply that

$$u(x_0, y_0) \leqq \tfrac{1}{4}[g_\nu(x_0 + h, y_0 + k) + g_\nu(x_0 - h, y_0 + k) + g_\nu(x_0 - h, y_0 - k) + g_\nu(x_0 + h, y_0 - k)]$$

for $h^2 + k^2 = r^2$. Subdivide $C(x_0, y_0; r)$ into n equal arcs by points $(x_0 + h_i, y_0 + k_i)$, $i = 1, \ldots, n$. Write the preceding inequality for h_i, k_i, $i = 1, \ldots, n$. After addition, it follows for $n \to \infty$ that $u(x_0, y_0) \leqq L(g_\nu; x_0, y_0; r)$. By 1.4 this implies that $L(u; x_0, y_0; r)$ exists and is $\geqq u(x_0, y_0)$. Hence u is subharmonic by 2.3.

As a special case of the preceding theorem, a function $u(x, y)$ is subharmonic if the surface $z = u(x, y)$ is convex in the sense that it is intersected in a convex curve by every plane which is parallel to the z-axis. Functions u with this property were the first ones to be considered as generalizations of convex functions of a single variable. Subharmonic functions with such special convexity properties were studied in detail by MONTEL [**3**] and by VALIRON [**1**].

3.23. If $u \geqq 0$ is subharmonic in G, then for $\alpha > 1$ the function $v = u^\alpha$ is also subharmonic, since t^α is increasing and convex if $\alpha > 1$ (cf. 3.13). More generally, if $u_1, \ldots, u_n$ are subharmonic and $\geqq 0$ in G and if $\alpha > 1$, then $u = (u_1^\alpha + \cdots + u_n^\alpha)^{1/\alpha}$ is also subharmonic. If $u_1, \ldots, u_n$ are of class $K^{(2)}$, then the relation $\Delta u \geqq 0$ reduces, by direct computation, to an inequality which is easily recognized as an immediate consequence of the inequality of SCHWARZ. In the general case the theorem follows then immediately by approximating $u_1, \ldots, u_n$ by integral means (see 2.21).

3.24. If u is of class PL in G (see 2.12), then it follows immediately from 2.12 that u^α is subharmonic for $\alpha > 0$. Conversely, if $u \geqq 0$ and $u \not\equiv 0$ in G and if u^α is subharmonic for every $\alpha > 0$, then u is of class PL (cf. MONTEL [**2**], p. 24). Indeed, $v_n = n(u^{1/n} - 1)$ is then clearly subharmonic for $n = 1, 2, \ldots$, and $v_n \searrow \log u$. Hence $\log u$ is subharmonic by 3.6.

3.25. We observed (see 2.8) that if u is subharmonic in G, then we have $A(u; x_0, y_0; r) \leqq L(u; x_0, y_0; r)$ whenever the circle $C(x_0, y_0; r)$ is comprised in G together with its interior. We shall consider now two theorems concerning the characterization of subharmonic functions in terms of inequalities involving only integral means. To simplify the statements we restrict ourselves to the case of continuous functions. We have then the theorem: if u is continuous in G, then u is subharmonic there if and only if $A(u; x_0, y_0; r) \leqq L(u; x_0, y_0; r)$ whenever the circle $C(x_0, y_0; r)$ is comprised in G together with its interior (BECKENBACH and RADÓ [**2**], p. 668). By 2.8 the condition is necessary. To prove its sufficiency, assume first that u is of class $K^{(2)}$. Let (x_0, y_0) be any point in G. The TAYLOR expansion yields then $L(u; x_0, y_0; \varrho) = u_0 + \frac{1}{4}\varrho^2(r_0 + t_0) + \sigma_1$, $A(u; x_0, y_0; \varrho) = u_0 + \frac{1}{8}\varrho^2(r_0 + t_0) + \sigma_2$, where u_0, r_0, t_0 are the values of u, u_{xx}, u_{yy} at (x_0, y_0), and $\sigma_i/\varrho^2 \to 0$ for $\varrho \to 0$, $i = 1, 2$. The inequality $A(u; x_0, y_0; \varrho) \leqq L(u; x_0, y_0; \varrho)$ implies that $r_0 + t_0 \geqq 8(\sigma_2 - \sigma_1)/\varrho^2$, and for $\varrho \to 0$ it follows that $r_0 + t_0 \geqq 0$. Hence u is subharmonic by 3.2. If u is only continuous, then the theorem follows by approximating u by integral means (see 2.21).

3.26. Similarly, if u is continuous and positive in G, then $\log u$ is subharmonic there if and only if $[A(u^2; x_0, y_0; \varrho)]^{1/2} \leqq L(u; x_0, y_0; \varrho)$ whenever the circle $C(x_0, y_0; \varrho)$ is comprised in G together with its interior (BECKENBACH and RADÓ [**2**], p. 665). The sufficiency of the condition is proved by the method used in 3.25. To prove the necessity, suppose that $\log u$ is subharmonic in G. Take any circle $C(x_0, y_0; \varrho)$ comprised in G together with its interior. Let h be the harmonic function in $C(x_0, y_0; \varrho)$ which coincides with $\log u$ on $C(x_0, y_0; \varrho)$. Denote by g the conjugate harmonic function, and put $f(z) = e^{h+ig}$, $z = x + iy$. According to a theorem of CARLEMAN [**1**], we have then $[A(|f|^2; x_0, y_0; \varrho)]^{1/2} \leqq L(|f|; x_0\, y_0; \varrho)$. Since $|f| = e^h$, we can write this inequality in the form $[A(e^{2h}; x_0, y_0; \varrho)]^{1/2} \leqq L(e^h; x_0, y_0; \varrho)$. As $\log u$ is subharmonic, it follows that

$$[A(u^2; x_0, y_0; \varrho)]^{1/2} \leqq [A(e^{2h}; x_0, y_0; \varrho)]^{1/2} \leqq L(e^h; x_0, y_0; \varrho) = L(u; x_0, y_0; \varrho).$$

3.27. The preceding proof depends upon analytic functions of a complex variable and therefore it does not apply in the case of three or more independent variables. It is not known at present whether an analogous theorem holds in the case of more than two variables. Clearly, the theorems of 3.25 and 3.26 are contributions to the problem of characterizing subharmonic properties in terms of conditions of the form

$$[A(u^\alpha; x_0, y_0; \varrho)]^{1/\alpha} \leqq [L(u^\beta; x_0, y_0; \varrho)]^{1/\beta}.$$

While the analogous problem for convex functions of a single variable was completely discussed (RADÓ [**3**]), the general problem for two or more variables seems to present serious difficulties.

3.28. If u is the limit of a sequence u_n of subharmonic functions, then u is also subharmonic if either $u_n \to u$ uniformly (see 3.3) or if $u_n \searrow u$ (see 3.6). MAZURKIEWICZ raised the problem of characterizing those functions which are limits of subharmonic functions in the sense of *convergence in the mean*. This problem was solved by SZPILRAJN [**1**], whose work we shall review presently.

3.29. A sequence of functions f_n converges *in the mean* to a function f in a domain G if for every domain G', comprised in G together with its boundary, the function f_n is defined and summable in G' for large n and $\iint_{G'} |f - f_n| \to 0$ for $n \to \infty$.

3.30. According to SZPILRAJN, a function u is almost subharmonic in a domain G if it satisfies the following conditions. a) u is summable on every measurable set completely interior to G. b) With the possible exception of a set of measure zero, we have $u(x_0, y_0) \leqq A(u; x_0, y_0; r)$ for every point (x_0, y_0) in G and for every r such that the circle $C(x_0, y_0; r)$ is comprised in G together with its interior.

3.31. If a sequence u_n of subharmonic functions converges in the mean to a function u in a domain G, then there exists in G a subharmonic function u^* such that $u^* = u$ almost everywhere in G. Proof. Take two domains G', G'' with boundaries B', B'' such that $G' + B' \subset G''$ and $G'' + B'' \subset G$. We have then an $r_0 > 0$ such that for every point (x_0, y_0) in G' and for $r < r_0$ the circle $C(x_0, y_0; r)$ is comprised in G'' together with its interior. Consider in G' the functions $A_r(x; y; u)$ and $A_r(x, y; u_n)$ (see 2.19). For fixed $r < r_0$ and for large n we have in G'

$$|A_r(x, y; u) - A_r(x, y; u_n)| \leqq \frac{1}{r^2 \pi} \iint_{\xi^2 + \eta^2 < r^2} |u(x+\xi, y+\eta) - u_n(x+\xi, y+\eta)| \, d\xi \, d\eta$$

$$\leqq \frac{1}{r^2 \pi} \iint_{G''} |u - u_n|.$$

The last integral converges to zero for fixed r and for $n \to \infty$, since u_n converges in the mean to u. Hence, for fixed $r < r_0$, $A_r(x, y; u_n)$ converges to $A_r(x, y; u)$ uniformly in G'. But $A_r(x, y; u_n)$ is subharmonic in G' (see 2.19). Hence (see 3.3) $A_r(x, y; u)$ is also subharmonic in G'. Finally, we have $A_{r_1}(x, y; u_n) \leqq A_{r_2}(x, y; u_n)$ for $r_1 < r_2$ by 2.9 and hence also $A_{r_1}(x, y; u) \leqq A_{r_2}(x, y; u)$. Consider now the sequence $u_k^*(x, y) = A_{1/k}(x, y; u)$. By what precedes, this sequence has the following properties. If G' is any domain completely interior to G, then for large k the functions u_k^*, u_{k+1}^*, ... are defined and subharmonic in G and $u_k^* \geqq u_{k+1}^* \geqq \cdots$ in G'. Hence, by 3.6, either $u_k^* \to -\infty$ in G or u_k^* converges in G to a subharmonic function u^*. On the other hand, $u_k^* \to u$ almost everywhere in G by a well-known theorem of LEBESGUE. It follows that $\lim u_k^* = u^*$ is subharmonic in G and $u^* = u$ almost everywhere in G.

3.32. If u is almost subharmonic in a domain G, then u is the limit of subharmonic functions in the sense of convergence in the mean. Proof. Consider again a pair of subdomains G', G'' as in 3.31. For small fixed r the function $A_r(x, y; u) = A(u; x, y; r)$ is then continuous in $G'' + B''$ and we have $u(x, y) \leqq A_r(x, y; u)$ almost everywhere in $G'' + B''$, by the definition of an almost subharmonic function. As a consequence, the theorem of TONELLI, referred to in 1.4, permits us to change the order of integrations necessary to show, starting with the inequality $u(x, y) \leqq A_r(x, y; u)$, that $A_r(x, y; u)$ is subharmonic. Finally, as it is well known, $A_r(x, y; u)$ converges in the mean to $u(x, y)$ for $r \to 0$ (see for instance MORREY [1], p. 687). Hence, if we put $u_n(x, y) = A_{1/n}(x, y; u)$, then the function u_n is subharmonic and converges in the mean to u, and the theorem is proved.

3.33. Combining 3.30, 3.31, 3.32 we obtain the following theorems (SZPILRAJN [1]). A function u is almost subharmonic in a domain G if and only if there exists in G a subharmonic function u^* such that $u = u^*$ almost everywhere in G. — A function u is almost subharmonic in a domain G if and only if it is the limit, in the sense of convergence in the mean, of some sequence of subharmonic functions.

3.34. Suppose that a function u satisfies in a domain G condition a) in 3.30 and that instead of condition b) in 3.30 it satisfies the following weaker condition b*): With the possible exception of a set of measure zero, there exists for every point (x_0, y_0) in G a $\varrho(x_0, y_0) > 0$ such that $u(x_0, y_0) \leqq A(u; x_0, y_0; r)$ for $r < \varrho(x_0, y_0)$. Then it does *not* follow that u is almost subharmonic in G (SZPILRAJN [1]). Example:

$$u(x, y) = \begin{cases} \log \dfrac{1}{x^2 + y^2} & \text{for} \quad x^2 + y^2 > 0. \\ 0 \quad \text{for} \quad x^2 + y^2 = 0. \end{cases}$$

3.35. We have the following corollary to 3.33. If u is almost subharmonic in a domain G, then there exists *exactly one* subharmonic function u^* such that $u = u^*$ almost everywhere in G (SZPILRAJN [1]). This follows immediately from 2.10.

3.36. It follows immediately from the definition of an almost subharmonic function (see 3.30) that a *continuous* almost subharmonic function is subharmonic (SZPILRAJN [1]).

3.37. As the reviewer could not find in the literature explicit applications of almost subharmonic functions, he takes the liberty to call attention to the following fact. Let u_n denote an *increasing* sequence of subharmonic functions in a domain G, such that there exists a finite constant M for which $\iint u_n < M$, $n = 1, 2, \ldots$, the integral being taken over G. By 1.4 the function $u = \lim u_n$ is then summable in G and condition b) in 3.30 is satisfied by u for every point of G. That is, *the limit of an increasing sequence of subharmonic functions is almost subharmonic as soon as it is summable.* In particular, there exists then a subharmonic function which differs from u at most at the points of a set of measure zero.

Various important problems lead to increasing sequences of subharmonic functions. We mention only the study of *the sweeping-out process* in Potential Theory (see for instance EVANS [4], part II) and the study of the convergence properties of power series of several complex variables (see MONTEL [2], pp. 56—60 and the remarks in RIESZ [4], p. 90 concerning HARTOGS [1]). It seems that the use of almost subharmonic functions might be of advantage in such cases. In a general way, the class of almost subharmonic functions presents the advantage of being closed under a considerably larger number of operations than the class of subharmonic functions.

Chapter IV.

Examples of subharmonic functions.

4.1. If $u(x, y)$ is a solution of a differential equation of the form $\Delta u = P$, where P is a function of $x, y, u, u_x, \ldots$ etc., then u is subharmonic in every domain G in which P is $\geqq 0$ (see 3.2). For various inferences from this remark see BRELOT ([1], pp. 52—55).

4.2. If h is harmonic in a domain G, then $-h$ is also harmonic there, and by 3.4 the functions $\overset{+}{h} = \overline{h, 0}$ and $|h| = \overline{h, -h}$ are subharmonic. For $\alpha \geqq 1$ the function $|h|^\alpha$ is also subharmonic by 3.23. More generally, if $h_1, \ldots, h_n$ are harmonic, then for $\alpha \geqq 1$ the function $u = (|h_1|^\alpha + \cdots + |h_n|^\alpha)^{1/\alpha}$ is subharmonic, on account of 3.23. If $f(z) = h_1 + i h_2$, $z = x + iy$, is an analytic function of the complex variable z, then $|f| = (h_1^2 + h_2^2)^{1/2}$ is subharmonic by the preceding remark.

4.3. As a matter of fact, $|f|$ is of class *PL* (see 2.12). This follows immediately from the remark that $\log|f|$ is harmonic at points where $f \neq 0$ and that the limit of $\log|f|$ is equal to $-\infty$ at points where $f = 0$. The far-reaching implications of the preceding facts were first emphasized by F. RIESZ [**1, 3, 4, 5**], whose papers contain also a number of historical references. Subsequent papers by various authors (practically all the papers quoted in this report) contain many important applications of subharmonic functions in the theory of analytic functions of a complex variable. MONTEL [**2**] and PRIVALOFF [**3, 4**] give particularly detailed presentations.

4.4. If $f = h_1 + i h_2$ is an analytic function of the complex variable $z = x + iy$, then h_1, h_2 are called conjugate harmonic functions. The CAUCHY-RIEMANN equations yield the relations $h_{1x}^2 + h_{2x}^2 = h_{1y}^2 + h_{2y}^2$, $h_{1x} h_{2x} + h_{1y} h_{2y} = 0$ for pairs of conjugate harmonic functions. As a generalization, three functions $u(x, y)$, $v(x, y)$, $w(x, y)$ are said to form *a triple of conjugate harmonic functions* (BECKENBACH and RADÓ [**1**]) if the following conditions are satisfied. 1) u, v, w are harmonic. 2) $E = G$, $F = 0$, where $E = u_x^2 + v_x^2 + w_x^2$, $G = u_y^2 + v_y^2 + w_y^2$, $F = u_x u_y + v_x v_y + w_x w_y$. According to a theorem of WEIERSTRASS, the surface represented by the equations $X = u(x, y)$, $Y = v(x, y)$, $Z = w(x, y)$ is then a minimal surface (X, Y, Z are CARTESian coordinates), and conversely every minimal surface can be represented in this form. As a generalization of 4.3 we have the following theorem (BECKENBACH and RADÓ [**1**], p. 653). If u, v, w form a triple of conjugate harmonic functions, then $(u^2 + v^2 + w^2)^{1/2}$ is of class *PL* (see 2.12). While the converse is false, it is true that if u, v, w are continuous in a domain G and if $[(u + a)^2 + (v + b)^2 + (w + c)^2]^{1/2}$ is of class *PL* for every choice of the constants a, b, c, then u, v, w form a triple of conjugate harmonic functions (BECKENBACH and RADÓ [**1**], p. 654). The first theorem follows by an elementary discussion of the explicit expression for $\Delta \log(u^2 + v^2 + w^2)^{1/2}$. To prove the second theorem, observe that since $[(u + a)^2 + (v + b)^2 + (w + c)^2]^{1/2}$ is of class *PL*, the function $f = (u + a)^2 + (v + b)^2 + (w + c)^2$ is subharmonic by 3.24. Let $C(x_0, y_0; r)$ be any circle comprised in G together with its interior. We have then $f(x_0, y_0) \leqq L(f; x_0, y_0; r)$. After some computation, this inequality leads to

$$0 \geqq u(x_0, y_0)^2 + v(x_0, y_0)^2 + w(x_0, y_0)^2 - L(u^2 + v^2 + w^2; x_0, y_0; r)$$
$$- 2a[L(u; x_0, y_0; r) - u(x_0, y_0] - \cdots - 2c[L(w; x_0, y_0; r) - w(x_0, y_0)].$$

Clearly, if a linear function of a, b, c has a constant sign, the coefficients of a, b, c must vanish. Hence $u(x_0, y_0) = L(u; x_0, y_0; r)$, and thus u is harmonic (by the so-called converse of GAUSS' theorem; see KELLOGG [**1**], p. 224 or combine 1.1 and 2.3). Similarly v and w are harmonic. Now that the derivatives of u, v, w are available, the relations $E = G$, $F = 0$ (cf. 4.4) can be proved by an elementary discussion of the inequality $\Delta \log f \geqq 0$.

4.5. Subharmonic functions are related to *surfaces of negative Gaussian curvature* as follows. Let a surface S be given, in terms of CARTESian coordinates X, Y, Z, by equations of the form $X = u(x, y)$, $Y = v(x, y)$, $Z = w(x, y)$. Put $E = u_x^2 + v_x^2 + w_x^2$, $G = u_y^2 + v_y^2 + w_y^2$, $F = u_x u_y + v_x v_y + w_x w_y$. Then the GAUSSian curvature K can be expressed, as it is well known, in terms of E, F, G. Suppose that $E = G$, $F = 0$ (that is, the surface S is given in terms of isothermic parameters). If we put $E = G = \lambda$, then the expression for K in terms of E, F, G reduces to $K = -(1/2\lambda)\, \Delta \log \lambda$. As $\lambda > 0$, it follows that $K \leqq 0$ if and only if $\Delta \log \lambda \geqq 0$, that is if and only if λ is of class PL (see 2.12). See BECKENBACH and RADÓ [**2**] for various geometrical consequences of this relationship between subharmonic functions and surfaces of negative curvature.

Suppose now only that λ is subharmonic. Then it does *not* follow that the GAUSSian curvature of the surface is $\leqq 0$. If however λ is subharmonic for *every* representation of the surface in terms of isothermic parameters, then the GAUSSian curvature of the surface is $\leqq 0$ (BECKENBACH [**1**]). This follows immediately from 2.13. This theorem has various interesting geometrical implications (see BECKENBACH [**1**]).

4.6. We shall consider presently subharmonic functions arising in Potential Theory (RIESZ [**4, 5**] and EVANS [**4**]). We shall use the general notion of a *positive mass-distribution* (RADON [**1**]). Let us first recall some properties of the class (B) of point-sets which are measurable in the sense of BOREL (BOREL [**1**], HAUSDORFF [**1**], KURATOWSKI [**1**]). The class (B) can be characterized as the *smallest* one of all classes K with the following properties. a) Every closed set belongs to K. b) If $S_1, S_2, \ldots, S_n, \ldots$ is a finite or infinite sequence of sets belonging to K, then $S_1 + S_2 + \cdots$ and $S_1 S_2 \cdots$ also belong to K. It follows easily that if a set S belongs to (B), then the complement of S (the set of points not in S) also belongs to (B). It is then immediate that the class (B) can be also characterized as the smallest one of all classes K^* with the following properties. α) K^* contains every set defined by two relations of the form $x_1 \leqq x < x_2$, $y_1 \leqq y < y_2$. β) If S_1, S_2 belong to K^*, then $S_1 S_2$ also belongs to K^*. γ) If S_2 and $S_1 \subset S_2$ belong to K^*, then $S_2 - S_1$ also belongs to K^*. δ) If $S_1, S_2, \ldots, S_n, \ldots$ is a finite or infinite sequence of *non-overlapping* sets belonging to K^*, then $S_1 + S_2 + \cdots$ also belongs to K^*. The equivalence of these two definitions of the class (B) has the following consequence. Denote by $(\mathfrak{P})$ the class of all sets which possess a certain property $\mathfrak{P}$. If it can be shown that $(\mathfrak{P})$ satisfies the conditions α), β), γ), δ), then we can assert that *every* set of class (B) possesses the property $\mathfrak{P}$.

4.7. *Positive mass-distributions.* In the sequel the letters E, e (with subscripts if necessary) will *always* refer to sets of class (B).

Given a bounded set E, let there be assigned to every subset e of E (including the empty set and also E itself) a finite real number $\mu(e)$ such that the following conditions are satisfied. 1) $\mu(e) \geqq 0$. 2) If $e_1, e_2, \ldots$ is a finite or infinite sequence of *non-overlapping* subsets of E, then $\mu(e_1 + e_2 + \cdots) = \mu(e_1) + \mu(e_2) + \cdots$. 3) $\mu(0) = 0$, where $\mu(0)$ denotes the number assigned to the empty set. These conditions being satisfied, $\mu(e)$ is called a positive mass-distribution on E. That is, a positive mass-distribution $\mu(e)$ is a non-negative (and hence monotonic) absolutely additive set-function in the sense of RADON [1]. The theory of these set-functions has been developed to a high degree of efficiency by RADON and it seems that they are generally accepted tools in dealing with problems in Potential Theory. We shall list presently a few facts concerning positive mass-distributions which will be used in the sequel.

4.8. Let E^* be a set containing the set E on which $\mu(e)$ is defined. Define, for subsets e^* of E^*, $\mu^*(e^*) = \mu(e^* E)$ (a product of two or more sets denotes the set of their common points). Clearly, $\mu^*(e^*)$ is a positive mass-distribution on E^*. If $e^* \subset E$, then $\mu^*(e^*) = \mu(e^*)$, and if $e^* E = 0$ then $\mu^*(e^*) = 0$. Roughly speaking, μ^* vanishes outside of E and μ^* is equal to μ on E. Using this remark, we can always assume that μ is defined on some set E of a convenient type (the interior of a large circle, for instance). RADON [1] assumes that μ is defined on an interval given by relations of the form $x_1 \leqq x < x_2$, $y_1 \leqq y < y_2$. While such assumptions simplify the presentation of the proofs, for the applications it is more convenient to state the theorems for a general set E of class (B).

4.9. If $e_1 \subset e_2$, then clearly $\mu(e_1) \leqq \mu(e_2)$.

4.10. If $e_1 \subset e_2 \subset \cdots$ and $e = e_1 + e_2 + \cdots$, then clearly $\mu(e_n) \to \mu(e)$.

4.11. Given a subset e of E, and an $\varepsilon > 0$, we have a *closed* subset e' of e such that $\mu(e) - \mu(e') < \varepsilon$ (RADON [1], pp. 1313—1314). That is, $\mu(e)$ is the least upper bound of $\mu(e')$ for all *closed* subsets e' of e. This fundamental property is proved on the basis of the remark at the end of 4.6.

4.12. Given a positive mass-distribution $\mu(e)$ on E, it might be possible to extend the definition of $\mu(e)$ to a class K^* of subsets of E in such a way that the properties 1), 2), 3) in 4.7 and also the property expressed by the theorem of 4.11 remain valid for all the sets of the class K^*. RADON [1] shows that in a certain sense there exists a largest class K^* satisfying these conditions, and he calls this class K^* the natural range of definition for $\mu(e)$. For instance, if $\mu(e)$ is the LEBESGUE measure of e, then the natural range of definition consists of all sets measurable in the sense of LEBESGUE. The theorem of 4.11 expresses the important fact that the natural range of definition always includes all sets of class (B). In other words, the class (B) is large enough to possess the closure properties listed in 4.6 and is also small enough to make valid the theorem of 4.11. *As stated in 4.7, we consider only sets of class* (B) *in connection with positive mass-distributions.*

4.13. Let $\mu_1(e)$ and $\mu_2(e)$ be given on an *open* set E. If $\mu_1(e)$ and $\mu_2(e)$ have the same value for every *open* subset of E, then $\mu_1 \equiv \mu_2$ on E. Observe that μ_1, μ_2 have then the same value for all *closed* subsets also, and apply 4.11.

4.14. Let us denote generally by s and b the set of the interior points and of the boundary points respectively of a square. Then we have the following corollary to 4.13. If $\mu_1(e)$ and $\mu_2(e)$ are given on an *open* set E; and if $\mu_1(s) = \mu_2(s)$ whenever $s + b$ is comprised in E, then $\mu_1 \equiv \mu_2$ on E. This may be seen as follows. Denote by $S(\xi)$ the set of those points of E which are located on the line $x = \xi$. If $\xi_1, \ldots, \xi_n$ are distinct, then we have $\mu_1(S(\xi_1)) + \cdots + \mu_1(S(\xi_n)) \leqq \mu(E)$. It follows immediately that $\mu_1(S(\xi)) = 0$, except possibly for a denumerable set of ξ-values. The same holds for $\mu_2(S(\xi))$, and we have a similar statement in terms of the y-coordinate. It follows that we have a point (x_0, y_0) with the following property. Denote by D_n the subdivision of the plane by means of the lines $x = x_0 + k/2^n$, $y = y_0 + j/2^n$, $k, j = 0, \pm 1, \ldots$ Let $s + b$ be any closed square of D_n. Then $\mu_1(E\,b) = \mu_2(E\,b) = 0$. On account of 4.9, 4.10 and 4.13 the theorem follows now immediately if we approximate the open subsets of E by squares taken from D_n.

4.15. *The Stieltjes-Radon integral* (Radon [1], p. 1322). Let $\mu(e)$ be a positive mass-distribution given on E. Denote by Q a variable point of E with coordinates (ξ, η). We shall write $Q = (\xi, \eta)$ in the sense that we shall use whichever of the notations Q and (ξ, η) will be more convenient. Let $f(Q) = f(\xi, \eta)$ be a function which is *uniformly continuous* on E. Subdivide E into a finite number of non-overlapping subsets $e_1, \ldots, e_n$. Denote by δ_k the diameter of e_k (that is, the least upper bound of the distances of pairs of points in e_k) and by δ the largest one of $\delta_1, \ldots, \delta_n$. We shall say that $e_1, \ldots, e_n$ form a subdivision D of E with norm δ. Pick a point $Q_k = (\xi_k, \eta_k)$ in e_k, $k = 1, \ldots, n$, and form the sum $\Sigma = \Sigma f(Q_k)\,\mu(e_k)$, $k = 1, \ldots, n$. In exactly the same way as in the case of the Riemann integral, it follows that the sum Σ approaches a limit, depending only upon f and μ, if the norm of the subdivision approaches zero. This limit is the Stieltjes-Radon integral $\int_E f(Q)\,d\mu(e_Q)$. The symbol e_Q is used to avoid misunderstandings in case f depends upon further variables.

4.16. Note that the STIELTJES-RADON integral is defined, for the time being, only for functions which are *uniformly contiuuous* on E. We shall consider later one of the various possible generalizations. It seems unnecessary to state explicitly all the simple properties of the STIELTJES-RADON integral which will be used in the sequel. As an example, we mention the following property. Suppose that E_1 is a subset of E, such that $\mu(E_1) = 0$. Put $E_2 = E - E_1$. Then $\int\limits_E f(Q)\, d\mu(e_Q) = \int\limits_{E_2} f(Q)\, d\mu(e_Q)$. This becomes obvious if we observe that we can subdivide E_1 and E_2 separately and that $\mu(E_1) = 0$ implies $\mu(e_1) = 0$ for every subset e_1 of E_1.

4.17. An important special case of positive mass-distributions is obtained as follows. Let $w(Q) = w(\xi, \eta)$ be a non-negative summable function on E. For convenience, we shall use notations like $\iint w(\xi, \eta)\, d\xi\, d\eta = \iint w(Q)\, da_Q$, where the symbol da_Q is to remind us of the variable of integration $Q = (\xi, \eta)$, of the area-element $d\xi\, d\eta$, and also of the fact that we are dealing with a LEBESGUE integral, in contradistinction with STIELTJES-RADON integrals. Consider now the set-function $\mu(e) = \iint\limits_e w(Q)\, da_Q$ on E. Obviously $\mu(e)$ is a positive mass-distribution on E. We have, for every function $f(Q)$ which is *uniformly continuous* on E, the relation $\int\limits_E f(Q)\, d\mu(e_Q) = \iint\limits_E f(Q)\, w(Q)\, da_Q$. To see this, take a subdivision $e_1, \ldots, e_n$ of E, and use the uniform continuity of f in comparing the sums $\Sigma f(Q_k)\, \mu(e_k)$ and

$$\Sigma \int\limits_{e_k} f(Q)\, w(Q)\, da_Q = \int\limits_E f(Q)\, w(Q)\, da_Q .$$

4.18. Given the positive mass-distribution $\mu(e)$ on E, we have to define the integral (potential of the negative mass-distribution $-\mu(e)$)

$$-\int\limits_E \log \frac{1}{PQ}\, d\mu(e_Q) .$$

The symbol PQ denotes the distance of the points $P = (x, y)$ and $Q = (\xi, \eta)$, where Q varies on E and P varies in the whole plane. The existence and the properties of this potential will be discussed presently (cf. RIESZ [5], part II; note that RIESZ uses a somewhat different definition of positive mass-distributions).

4.19. Let us put

$$l(P, Q) = l(x, y;\, \xi, \eta) = \begin{cases} -\log(1/PQ) & \text{for } P \neq Q, \\ -\infty & \text{for } P = Q. \end{cases}$$

For fixed Q, $l(P, Q)$ is clearly a subharmonic function of P and conversely. For fixed Q and $P \neq Q$, $l(P, Q)$ is a harmonic function of P and conversely.

4.20. Put, for $\sigma > 0$,

$$l^{(\sigma)}(P,Q) = l^{(\sigma)}(x,y;\xi,\eta) = \begin{cases} l(P,Q) & \text{for } PQ \geqq \sigma, \\ -\log(1/\sigma) & \text{for } PQ \leqq \sigma. \end{cases}$$

For fixed Q, $l^{(\sigma)}(P,Q)$ is a subharmonic function of P by 3.4, and conversely. Clearly, $l^{(\sigma)}(P,Q)$ is a continuous function of P and Q, and its continuity is uniform if P and Q vary on bounded sets. Also $l^{(\sigma)}(P,Q) \searrow l(P,Q)$ for $\sigma \searrow 0$.

4.21. Put, for $r > 0$,

$$l_r(P,Q) = l_r(x,y;\xi,\eta) = \frac{1}{2\pi}\int_0^{2\pi} l(x + r\cos\varphi,\ y + r\sin\varphi;\ \xi,\ \eta)\,d\varphi.$$

We find by direct elementary computation the formula $l_r(P,Q) = l^{(r)}(P,Q)$ (see 4.20).

4.22. Put, for $r > 0$ and $\sigma > 0$,

$$l_r^{(\sigma)}(P,Q) = l_r^{(\sigma)}(x,y;\xi,\eta) = \frac{1}{2\pi}\int_0^{2\pi} l^{(\sigma)}(x + r\cos\varphi,\ y + r\sin\varphi;\ \xi,\eta)\,d\varphi.$$

By 4.20 and 4.21 we have then $l_r^{(\sigma)}(P,Q) \searrow l_r(P,Q)$ for $\sigma \searrow 0$. Since $l_r^{(\sigma)}(P,Q)$ and $l_r(P,Q)$ are continuous, it follows by a well-known theorem of DINI (see for instance PÓLYA-SZEGŐ [**1**], p. 225, problem 126) that $l_r^{(\sigma)}(P,Q) \to l_r(P,Q)$ for $\sigma \to 0$ *uniformly* if P and Q vary on bounded sets (since such sets are comprised in bounded *closed* sets). By 4.19 to 4.21 we have for $r + \sigma < PQ$ the formula $l_r^{(\sigma)}(P,Q) = l(P,Q)$.

4.23. We define now (cf. RIESZ [**5**], part II and DANIELL [**1**])

$$u(P) = \int_E l(P,Q)\,d\mu(e_Q) = -\int_E \log\frac{1}{PQ}\,d\mu(e_Q) = \lim_{\sigma\to 0} u^{(\sigma)}(P),$$

where, for $\sigma > 0$,

$$u^{(\sigma)}(P) = \int_E l^{(\sigma)}(P,Q)\,d\mu(e_Q).$$

Since $l^{(\sigma)}(P,Q)$ is uniformly continuous if P and Q vary on bounded sets, $u^{(\sigma)}(P)$ is a well-defined and continuous function of P in the whole plane. From $\mu \geqq 0$ it follows that $u^{(\sigma)}(P)$ decreases if $\sigma > 0$ decreases, and thus $u(P) = \lim u^{(\sigma)}(P)$ exists for every P, the value of $u(P)$ being possibly equal to $-\infty$. We have $-\infty \leqq u(P) < +\infty$, $u^{(\sigma)}(P) \searrow u(P)$ for $\sigma \searrow 0$, and we can assert also that $u(P)$ is upper semi-continuous, since we obtained $u(P)$ as the limit of a decreasing sequence of continuous functions. We shall see now that $u(P)$ is subharmonic.

4.24. Put, for $\sigma > 0$ and $r > 0$,

$$u_r^{(\sigma)}(P) = u_r^{(\sigma)}(x,y) = \frac{1}{2\pi}\int_0^{2\pi} u^{(\sigma)}(x + r\cos\varphi,\ y + r\sin\varphi)\,d\varphi.$$

Since $l^{(\sigma)}(P, Q)$ is continuous (see 4. 20, 4. 23), we obtain by an obviously permissible change in the order of integrations the formula $u_r^{(\sigma)}(P) = \int_E l_r^{(\sigma)}(P, Q)\, d\mu(e_Q)$. On account of 4. 22, $l_r^{(\sigma)}(P, Q) \to l_r(P, Q)$ uniformly for $\sigma \to 0$. Hence $u_r^{(\sigma)}(P) \to \int_E l_r(P, Q)\, d\mu(e_Q)$ for $\sigma \to 0$.

4. 25. Since $l^{(\sigma)}(P, Q)$ is a subharmonic function of P (see 4. 20) we have, by 2. 3, $l_r^{(\sigma)}(P, Q) \geqq l^{(\sigma)}(P, Q)$. Hence (see 4. 24), $u_r^{(\sigma)}(P) \geqq u^{(\sigma)}(P)$. Thus $u^{(\sigma)}(P)$ is subharmonic by 2. 3. Suppose now that the point P has a positive distance δ from the set E. For $\sigma < \delta/2$, $r < \delta/2$ we have then by 4. 22 and 4. 24

$$u_r^{(\sigma)}(P) = \int_E l_r^{(\sigma)}(P, Q)\, d\mu(e_Q) = \int_E l(P, Q)\, d\mu(e_Q) = \int_E l^{(\sigma)}(P, Q)\, d\mu(e_Q) = u^{(\sigma)}(P).$$

Consider then an open set O such that $PQ > \delta > 0$ for P in O and Q in E. By what precedes, we have for P in O, $r < \delta/2$, $\sigma < \delta/2$ the relations $u_r^{(\sigma)}(P) = u^{(\sigma)}(P)$ and $u^{(\sigma)}(P) = \int_E l(P, Q)\, d\mu(e_Q)$. The first relation shows that $u^{(\sigma)}(P)$ is harmonic in O (converse of GAUSS' theorem, see KELLOGG [1], p. 224). The second relation shows that $u(P) = \lim u^{(\sigma)}(P) = u^{(\sigma)}(P)$ in O for $\sigma < \delta/2$. Thus $u(P)$ is also harmonic in O.

4. 26. As $u(P)$ is the limit of a decreasing sequence of subharmonic functions $u^{(\sigma)}(P)$, it follows by 3. 6 that either $u \equiv -\infty$ in the whole plane or u is subharmonic in the whole plane. The first case is excluded by the remark that u is harmonic outside of a sufficiently large circle (see 4. 25). Consequently *the potential*

$$u(P) = -\int_E \log \frac{1}{PQ}\, d\mu(e_Q)$$

is subharmonic in the whole plane (RIESZ [5], part II). In particular (see 1. 10), $u > -\infty$ almost everywhere.

4. 27. Let E_1 be an *open* subset of E, such that $\mu(E_1) = 0$. Put $E_2 = E - E_1$. We have then (see 4. 16)

$$u^{(\sigma)}(P) = \int_E l^{(\sigma)}(P, Q)\, d\mu(e_Q) = \int_{E_2} l^{(\sigma)}(P, Q)\, d\mu(e_Q),$$

and consequently (see 4. 23)

$$u(P) = \int_{E_2} l(P, Q)\, d\mu(e_Q).$$

Hence, on account of the remark at the end of 4. 25, $u(P)$ is harmonic in E_1. Summing up: *the potential $u(P)$ is harmonic on every open set which contains no mass.*

4.28. As $u^{(\sigma)} \searrow u$ for $\sigma \searrow 0$, we have (see 1.4, 4.24, 4.21, 4.23)

$$L(u; x, y; r) = \lim_{\sigma\to 0} L(u^{(\sigma)}; x, y; r) = \lim_{\sigma\to 0} u_r^{(\sigma)}(P) = \int_E l_r(P, Q)\, d\mu(e_Q)$$

$$= \int_E l^{(r)}(P, Q)\, d\mu(e_Q) = u^{(r)}(x, y)\,.$$

These formulas throw a new light on the definition of the potential $u(P)$ given in 4.23.

4.29. Take a point $P_0 = (x_0, y_0)$ and choose r so large that the set E is completely interior to the circle $C(x_0, y_0; r)$. For Q in E we have then $P_0Q < r$ and hence $l_r(P_0, Q) = \log r$ (see 4.21). It follows then by 4.28 that $\mu(E)\log r = L(u; x_0, y_0; r)$.

4.30. On $C(x_0, y_0; r)$ we can use the derivatives of u, since u is harmonic there by 4.27. Let us write C_r for $C(x_0, y_0; r)$. We have then (on account of 4.29) the formula

$$\frac{1}{2\pi}\int_{C_r} \frac{\partial u}{\partial n_e}\, ds = r\frac{d}{dr} L(u; x_0, y_0; r) = \mu(E)\,.$$

4.31. Consider now any smooth JORDAN curve Γ, such that the set E is completely interior to Γ. We can choose then the circle $C(x_0, y_0; r)$ of 4.29 in such a way that Γ is enclosed by $C(x_0, y_0; r)$. As u is harmonic between and on these two curves, the line integral of 4.30 has the same value for both curves. Hence, the total mass $\mu(E)$ can be expressed in terms of the potential u by the familiar formula

$$\mu(E) = \frac{1}{2\pi}\int_\Gamma \frac{\partial u}{\partial n_e}\, ds\,,$$

where Γ is any smooth JORDAN curve such that E is completely interior to Γ (RIESZ [5], part II).

4.32. The problem of expressing the mass $\mu(e)$, for every subset of class (B), in terms of the potential u will be considered in Chapter V.

4.33. Let $w(Q) = w(\xi, \eta)$ be a non-negative summable function on a bounded set E of class (B). Consider the function (see 4.17 for notations)

$$v(P) = -\iint_E \log\frac{1}{PQ}\, w(Q)\, da_Q\,.$$

This integral can be interpreted in various ways. One usual interpretation is expressed by the formula (cf. 4.20)

$$v(P) = \lim_{\sigma\to 0}\iint_E l^{(\sigma)}(P, Q)\, w(Q)\, da_Q\,.$$

By 4.17 and 4.23 it follows that

$$v(P) = -\int_E \log\frac{1}{PQ}\, d\mu(e_Q)\,,$$

where the positive mass-distribution $\mu(e)$ is defined by $\mu(e) = \int_e w(T)\, d\alpha_T$.

That is, the potentials of mass-distributions with a summable negative density $-w(\xi, \eta)$, as considered in Potential Theory, are included among the potentials of general negative mass-distributions $-\mu(e)$.

4.34. Using the fact that potentials of negative mass-distributions are subharmonic, it is possible to construct subharmonic functions with various types of discontinuities (Riesz [**5**], part I, p. 336; Brelot [**1**], pp. 42—47; Evans [**5**], p. 421). These examples show the great variety of new possibilities as compared with the one-dimensional case of convex functions of a single variable.

Chapter V.

Harmonic majorants of subharmonic functions.

5.1. Throughout this Chapter, u will denote a function which is subharmonic in a domain G. Consider a region $G' + B'$ comprised in G and a function H which is continuous in $G' + B'$ and harmonic in G'. If $H \geqq u$ on B', then $H \geqq u$ in G' also, by the definition of a subharmonic function. Naturally, one will try to use a harmonic majorant H which is as small as possible. Suppose that $G' + B'$ is a Dirichlet region and suppose also that u is *continuous*. The solution H of the Dirichlet problem for G' with the boundary condition $H = u$ on B' is then obviously the best harmonic majorant in G'. If however u is not continuous, then it is not clear that there exists a harmonic majorant in G' which should be considered the best. This situation lead to investigations which will be reviewed presently.

5.2. Consider a Dirichlet region $G' + B'$ comprised in G. By 1.3 we have on B' a sequence of continuous functions φ_k such that $\varphi_k \searrow u$ on B'. Denote by H_k the solution of the Dirichlet problem for G' with the boundary condition $H_k = \varphi_k$ on B'. Then we have (see 1.3) $H_k \geqq H_{k+1}$ and $H_k \geqq u$ on $G' + B'$, and therefore H_k converges in G' to a harmonic function $\bar{h} \geqq u$.

5.3. The function $\bar{h}$ of 5.2 has the following property. Let H be continuous and $\geqq u$ in $G' + B'$ and harmonic in G'. Then $H \geqq \bar{h}$ in G' (Riesz [**5**], part I, p. 334). To see this, give any $\varepsilon > 0$. As $\varphi_k \searrow u < H + \varepsilon$ and as φ_k and H are continuous on the closed set B', it follows by the Heine-Borel theorem that we have a $\varkappa = \varkappa(\varepsilon)$ such that $\varphi_k < H + \varepsilon$ on B' for $k > \varkappa$. But $H_k = \varphi_k$ on B' and H_k is harmonic in G'. Hence $H_k < H + \varepsilon$ for $k > \varkappa$. As $\varepsilon > 0$ is arbitrary and $H_k \searrow \bar{h}$ in G', it follows that $h \leqq H$ in G'.

5.4. The function $\bar{h}$ of 5.2 depends only upon the values of u on B' (RIESZ [5], part I, pp. 333—334). Indeed, take a second sequence φ'_k and denote by H'_k and $\bar{h}'$ the corresponding harmonic functions. By 5.3 we have $\bar{h} \leqq H'_k$ and $\bar{h}' \leqq H_k$ in G' and the theorem follows for $k \to \infty$.

Consider now two functions u_1, u_2 which are subharmonic in G and equal to each other on B'. As we can use then the same sequence φ_k for both functions, there corresponds the same function $\bar{h}$ to u_1 and to u_2.

The harmonic function $\bar{h}$ defined in 5.2 will be called *the best harmonic majorant* (B. H. M.) of u in G' (RIESZ [5], part I, p. 334). By what precedes, u depends solely upon the values of u on the boundary B' of G'. The term best harmonic majorant suggests various questions which will be considered later in this chapter. It should be noted that a B. H. M. is only defined for DIRICHLET subregions $G' + B'$. If u is continuous on B', then we can use $\varphi_k = u$ as the sequence leading to h, and it follows that in this special case $\bar{h}$ is simply the solution of the DIRICHLET problem for G' with the boundary condition $\bar{h} = u$ on B'. Another important special case arises if $G' + B'$ is a closed circular disc, while u is a general subharmonic function. The function H_k of 5.2 is then given in G' by the formula of POISSON. As the POISSON kernel is positive and $\varphi_k \searrow u$ on B', we infer from 1.4 that $\bar{h}$ is also given by the formula of POISSON with u itself as the given boundary function. Clearly, a similar remark holds for DIRICHLET subregions with smooth boundaries.

5.5. Consider in G three JORDAN curves C_1, C_2, C_3, each of which is enclosed by the next one to the right, such that the three doubly connected domains D_{12}, D_{13}, D_{23} bounded by these curves are also comprised in G. Denote by $\bar{h}_{12}, \bar{h}_{13}, \bar{h}_{23}$ the B. H. M. of u in D_{12}, D_{13}, D_{23} respectively. Then $\bar{h}_{13} - \bar{h}_{12}$ is non-negative in D_{12} and vanishes continuously on C_1, and $\bar{h}_{13} - \bar{h}_{23}$ is non-negative in D_{23} and vanishes continuously on C_3 (RIESZ [5], part I, p. 341). Proof. Consider $\bar{h}_{13} - \bar{h}_{12}$, for instance. Take a sequence of continuous functions φ^i_k on C_i such that $\varphi^i_k \searrow u$ on C_i, $i = 1, 2, 3$. Denote by H^{13}_k the solution of the DIRICHLET problem for D_{13} with the boundary condition $H^{13}_k = \varphi^1_k$ on C_1, $H^{13}_k = \varphi^3_k$ on C_3, and let H^{12}_k have the same meaning with respect to D_{12}. Finally, denote by C_4 an auxiliary JORDAN curve in D_{12} which encloses C_1. Then the sequence $H_k = H^{13}_k - H^{12}_k$ converges uniformly on the boundary of the domain D_{14} bounded by C_1 and C_4 and hence (see KELLOGG [1], p. 248) this sequence converges *uniformly* in $D_{14} + C_1 + C_4$ to a limit function h which is continuous in $D_{14} + C_1 + C_4$, equal to zero on C_1, and equal to $\bar{h}_{13} - \bar{h}_{12}$ in D_{14}. This proves that $\bar{h}_{13} - \bar{h}_{12}$ vanishes continuously on C_1. By 5.3 we have $H^{13}_k \geqq \bar{h}_{12}$ in D_{12}, and for $k \to \infty$ it follows that $\bar{h}_{13} \geqq \bar{h}_{12}$ in D_{12}.

If the interior of C_2 is comprised in G, and if we denote be $\bar{h}_2$ the B. H. M. of u in the interior of C_2, then the same reasoning shows that $\bar{h}_2 - \bar{h}_{12}$ is non-negative in D_{12} and vanishes continuously on C_2.

5.6. Let G' be a bounded domain with boundary B'. Using subdivisions of the plane into congruent squares in a familiar fashion (KELLOGG [**1**], p. 317), we obtain a sequence of regions $G'_n + B'_n$ which approximate G' in the following sense. a) $G'_n + B'_n \subset G'$. b) $G'_n + B'_n \subset G'_{n+1}$. c) For every *closed* set S in G' there exists an $n_0 = n_0(S)$ such that S is in G'_n for $n > n_0$. d) B'_n consists of a finite number of JORDAN curves as smooth as desired (in particular, $G'_n + B'_n$ is a DIRICHLET region). The following statements are easy consequences of the preceding properties. e) Given $\varepsilon > 0$, denote by S_ε the set of those points in G' whose distance from B' is less than ε. Then for every $\varepsilon > 0$ there exists an $m = m(\varepsilon)$ such that B'_n is comprised in S_ε for $n > m$. f) The area of G'_n converges to the area of G, and consequently the measure of $G' - G'_n$ converges to zero.

5.7. Given a subharmonic function u in a domain G, consider a domain G' comprised in G. Suppose that there exists a function H_0 which is harmonic and $\geqq u$ in G' (this assumption is clearly satisfied if the boundary of G' is also comprised in G). Then there exists in G' a harmonic function h^* such that 1) $u \leqq h^*$ in G' and 2) every function H which is harmonic and $\geqq u$ in G' is also $\geqq h^*$ (RIESZ [**5**], part II, p. 358). Proof. Approximate G' by a sequence $G'_n + B'_n$ as described in 5.6. Denote by $\bar{h}_n$ the B. H. M. of u in G'_n. By 5.3 we have $u \leqq \bar{h}_n \leqq \bar{h}_{n+1} \leqq H_0$ in G'_n. It follows then from the theorem of HARNACK that the sequence $\bar{h}_n$ converges in G' to a function h^* which is harmonic in G' and which satisfies there the inequalities $u \leqq h^* \leqq H_0$. By 5.3 we have also $\bar{h}_n \leqq H$ for every function H which is harmonic and $\geqq u$ in G' and for $n \to \infty$ it follows that $h^* \leqq H$ in G'.

5.8. The harmonic function h^* of 5.7 is obviously unique. It may be called *the least harmonic majorant* (L. H. M.) of u in G' (RIESZ [**5**], part II, p. 357). If $G' + B'$ is a DIRICHLET region comprised in G, then the B. H. M. $\bar{h}$ and the L. H. M. h^* of u in G' both exist. Clearly $h^* \leqq \bar{h}$. As $\bar{h}$ depends solely upon the values of u on B' and h^* depends solely upon the values of u in G', it is not evident that $\bar{h}$ and h^* should be identical. The identity of $\bar{h}$ and h^* was established for special types of subregions $G' + B'$ by F. RIESZ ([**5**], part I, p. 334) and by BRELOT ([**1**], p. 18). We shall see later in this Chapter that $\bar{h}$ and h^* are always identical.

5.9. The majorants $\bar{h}$ and h^* depend upon u and upon G'. Some aspects of this dependence were investigated by MALCHAIR [**2**]. In the way of illustration we quote one of his results. Consider in a domain G a uniformly convergent sequence of subharmonic functions u_n.

Then the limit function u is also subharmonic by 3.3. Denote by $G'_n + B'_n$ a sequence of regions which approximate G in the sense of 5.6, and by $\bar{h}_n$ the B. H. M. of u_n in G'_n. Then $\bar{h}_n$ converges to the L. H. M. of u in G provided that this L. H. M. exists. The proof is similar to that in 5.7.

5.10. *A lemma on harmonic functions* (RIESZ [**5**], part I, p. 341). Consider two JORDAN curves C_1, C_2 such that C_1 is enclosed by C_2, and denote by D the doubly connected domain bounded by these curves. Let h be a function which is continuous on $D + C_1 + C_2$ and harmonic and *non-negative* in D. Take a smooth JORDAN curve Γ in D which encloses C_1. If $h = 0$ on C_1, then $\int_\Gamma (\partial h/\partial n_e)\, ds \geqq 0$, and if $h = 0$ on C_2, then $\int_\Gamma (\partial h/\partial n_e)\, ds \leqq 0$. Proof. *Special case.* Suppose that $h = 0$ on C_1, for instance, and suppose that C_1 is sufficiently *smooth.* Then the first and second derivatives of h remain continuous on C_1, and the line integral has the same value for Γ and for C_1 (KELLOGG [**1**], p. 212). The integral taken on C_1 is however obviously $\geqq 0$. If h vanishes on C_2, and if C_2 is sufficiently smooth, then the theorem is equally obvious. *General case.* Suppose that $h = 0$ on C_1, for instance. Take two *smooth* JORDAN curves C_3, C_4 such that each of the curves $C_3, C_1, \Gamma, C_4, C_2$ is enclosed by the next one to the right (C_3 being close to C_1 and C_4 close to C_2). Denote by H_{34} the solution of the DIRICHLET problem for the domain bounded by C_3 and C_4 with the boundary condition $H_{34} = 0$ on C_3, $H_{34} = h$ on C_4. Apply the *special case* of the theorem to $H_{34} - h$ in the domain between C_1 and C_4, then to H_{34} in the domain between C_3 and C_4, and combine the resulting inequalities. The case when $h = 0$ on C_2 is discussed in a similar manner.

5.11. F. RIESZ ([**5**], part I) introduced the following quantities in the study of subharmonic functions. Let u be subharmonic in a domain G. Take in G a pair of JORDAN curves C_1, C_2 such that C_1 is enclosed by C_2 and the domain D_{12} between C_1 and C_2 is comprised in G. Denote by $\bar{h}_{12}$ the B. H. M. of u in D_{12} and put

$$F(C_1; C_2; u) = \frac{1}{2\pi} \int_\Gamma \frac{\partial \bar{h}_{12}}{\partial n_e}\, ds,$$

where Γ is a smooth JORDAN curve in D_{12} which encloses C_1, and n_e refers to the outward normal of Γ. The quantity $F(C_1, C_2; u)$ is clearly independent of Γ (see KELLOGG [**1**], p. 212). If u_1, u_2 are both subharmonic in G, then clearly

$$F(C_1, C_2; u_1) + F(C_1, C_2; u_2) = F(C_1, C_2; u_1 + u_2).$$

5.12. If the interior of C_2 is comprised in G, then $F(C_1, C_2; u) \geqq 0$ (RIESZ [**5**], part I, p. 342). Proof. Denote by D_{12} the domain between C_1 and C_2 and by D_2 the interior of C_2. Let $\bar{h}_{12}$ and $\bar{h}_2$ be the B. H. M. of u in D_{12} and D_2 respectively. Take a smooth JORDAN curve Γ in D_{12} which encloses C_1. Then

$$\frac{1}{2\pi}\int_{\Gamma}\frac{\partial \bar{h}_2}{\partial n_e}\,ds = 0\,,\quad \frac{1}{2\pi}\int_{\Gamma}\frac{\partial \bar{h}_{12}}{\partial n_e}\,ds = F(C_1, C_2; u)\,,$$

and hence, by 5.5 and 5.10,

$$F(C_1, C_2; u) = -\frac{1}{2\pi}\int_{\Gamma}\frac{\partial(\bar{h}_2 - \bar{h}_{12})}{\partial n_e}\,ds \geqq 0\,.$$

5.13. Take in G three JORDAN curves C_1, C_2, C_3 such that C_1 is enclosed by C_2, C_2 is enclosed by C_3, and the domains D_{12}, D_{13}, D_{23} bounded by these curves are comprised in G. Then $F(C_1, C_2; u) \leqq F(C_1, C_3; u) \leqq F(C_2, C_3; u)$ (RIESZ [**5**], part I, p. 340). Proof. Take a smooth JORDAN curve Γ in D_{12} which encloses C_1 and denote by $\bar{h}_{12}, \bar{h}_{13}, \bar{h}_{23}$ the B. H. M. of u in D_{12}, D_{13}, D_{23} respectively. Then

$$F(C_1, C_3; u) - F(C_1, C_2; u) = \frac{1}{2\pi}\int_{\Gamma}\frac{\partial(\bar{h}_{13} - \bar{h}_{12})}{\partial n_e}\,ds\,,$$

and this integral is $\geqq 0$ by 5.5 and 5.10. The inequality $F(C_1, C_3; u) \leqq F(C_2, C_3; u)$ is proved in a similar way.

5.14. Consider the particular case when C_1 and C_2 are concentric circles with centre (x_0, y_0) and radii r_1 and $r_2 > r_1$. Denote by $\bar{h}_{12}$ the B. H. M. of u in the domain D_{12} between C_1 and C_2 and by C_r the concentric circle with radius r, $r_1 < r < r_2$. Then (see 1.5 for notations)

$$F(C_1, C_2; u) = r\frac{d}{dr}L(\bar{h}_{12}; x_0, y_0; r)\,.$$

On the other hand, the reasoning used in 1.12 and 1.13 shows that $L(\bar{h}_{12}; x_0, y_0; r) = a\log r + b$, where $a\log r + b$ is the linear function of $\log r$ which is equal to $L(u; x_0, y_0; r_1)$ for $r = r_1$ and to $L(u; x_0, y_0; r_2)$ for $r = r_2$. Combining these relations, we obtain the formula (RIESZ [**5**], part I, p. 340)

$$F(C_1, C_2; u) = \frac{L(u; x_0, y_0; r_2) - L(u; x_0, y_0; r_1)}{\log r_2 - \log r_1}\,.$$

5.15. The theorems of 2.4 and 2.5 appear now, on account of 5.14, as special cases of the theorems of 5.12 and 5.13.

5.16. Given a potential (cf. 4.23)

$$u(P) = -\int_{G}\log\frac{1}{PQ}\,d\mu(e_Q)\,,$$

where G is a bounded domain, there arises the problem to express the positive mass-distribution $\mu(e)$ in terms of u. If the distribution $\mu(e)$ is smooth, and if $G' + B'$ is a region in G with smooth boundary B', then the problem is solved by the classical formula (KELLOGG [1], pp. 155—156)

$$\mu(G') = \frac{1}{2\pi}\iint\limits_{G'} \Delta u(x,y)\,dx\,dy = \frac{1}{2\pi}\int\limits_{B'} \frac{\partial u}{\partial n_e}\,ds\,,$$

where n_e refers to the exterior normal with respect to G'. For a general $\mu(e)$ the problem was solved by G. C. EVANS in terms of a certain function of curves (EVANS [1], p. 271 and p. 285). We shall discuss this problem presently in terms of the quantities $F(C_1, C_2; u)$ introduced by F. RIESZ.

5. 17. The potential $u(P)$ of 5. 16 is subharmonic in the whole plane (see 4. 26) and therefore the preceding theorems apply to $u(P)$. Take two JORDAN curves C_1, C_2, such that C_1 is enclosed by C_2, and G is completely interior to C_1. By 4. 27 the potential u is harmonic on and between C_1 and C_2 and hence u is its own best harmonic majorant in the domain between C_1 and C_2. By 4. 31 and 5. 11 we obtain therefore for the total mass $\mu(G)$ the formula $\mu(G) = F(C_1, C_2; u)$. Consider next two JORDAN curves C_1, C_2 such that C_1 is enclosed by C_2 and both curves are comprised in a simply connected subdomain G' of G with $\mu(G') = 0$. Then, by 4. 27, u is harmonic in G' and again u is its own harmonic majorant in the domain between C_1 and C_2. If Γ is a smooth JORDAN curve which encloses C_1 and is enclosed by C_2, then it follows from the preceding remark that

$$F(C_1, C_2; u) = \frac{1}{2\pi}\int\limits_{\Gamma} \frac{\partial u}{\partial n_e}\,ds = 0\,,$$

since u is harmonic in and on Γ (see KELLOGG [1], p. 212).

5. 18. Take now five JORDAN curves $C_1, \ldots, C_5$ such that each one is enclosed by the next one to the right. Denote by D_i the interior of C_i, by D_{ij} the domain between C_i and C_j, and by $\bar{h}_i$, $\bar{h}_{ij}$ the B. H. M. of u in D_i, D_{ij} respectively. We have then $F(C_1, C_2; u) \leqq \mu(GD_3) \leqq F(C_4, C_5; u)$ (RIESZ [5], part II, pp. 331—335).

5. 19. To prove the preceding theorem, introduce on G the distributions $\mu'(e) = \mu(eGD_3)$, $\mu''(e) = \mu(e(G - GD_3))$, and the corresponding potentials (cf. 4. 16)

$$u'(P) = -\int\limits_{G} \log\frac{1}{PQ}\,d\mu'(e_Q) = -\int\limits_{GD_3} \log\frac{1}{PQ}\,d\mu(e_Q)\,,$$

$$u''(P) = -\int\limits_{G} \log\frac{1}{PQ}\,d\mu''(e_Q) = -\int\limits_{G-GD_3} \log\frac{1}{PQ}\,d\mu(e_Q)\,.$$

Then $\mu'(e) + \mu''(e) = \mu(e)$ and consequently $u'(P) + u''(P) = u(P)$. Hence (see 5. 11) $F(C_1, C_2; u') + F(C_1, C_2; u'') = F(C_1, C_2; u)$. We have, by 5. 17, $F(C_1, C_2; u'') = 0$ and $F(C_4, C_5; u') = \mu'(GD_3) = \mu(GD_3)$. Repeated application of 5. 13 yields $F(C_1, C_2; u') \leqq F(C_4, C_5; u')$. The inequality $F(C_1, C_2; u) \leqq \mu(GD_3)$ follows by combining these relations. The inequality $\mu(GD_3) \leqq F(C_4, C_5; u)$ is obtained in a similar fashion.

5.20. Consider now a *simply connected* domain G' comprised in G. Take a sequence of pairs of JORDAN curves C'_n, C''_n in G' such that 1) C'_n is enclosed by C''_n and 2) every closed set in G' is comprised in the interior of C'_n for sufficiently large n. Then $F(C'_n, C''_n; u) \to \mu(G')$ (RIESZ [**5**], part II, p. 336). This follows immediately from 5.18, 5.13 and 4.10.

5.21. By 5.20, $\mu(G')$ is determined in terms of u whenever G' is a simply connected subdomain of G. A similar reasoning yields the determination of $\mu(e)$ for multiply connected subdomains (RIESZ [**5**], part II, p. 336).

5.22. From 5.20 and 4.14 we infer the following theorem. If $\mu_1(e)$, $\mu_2(e)$ are positive mass-distributions on a bounded domain G, and if the coriesponding potentials

$$u_1(P) = -\int_G \log\frac{1}{PQ}\, d\mu_1(e_Q), \qquad u_2(P) = -\int_G \log\frac{1}{PQ}\, d\mu_2(e_Q)$$

are equal to each other in G, then $\mu_1(e) \equiv \mu_2(e)$ (remember that we consider only subsets e of class (B)). For the sake of accuracy it should be observed that F. RIESZ ([**5**], part II) considers positive mass-distributions defined in a somewhat different manner. In particular, his $\mu(e)$ is defined only for *open* sets e. The remark that the results of F. RIESZ include the preceding uniqueness theorem is due to EVANS ([**4**], part II, p. 203).

5.23. *A lemma on sequences of harmonic functions* (KELLOGG [**1**], Chapter XI). Let $G' + B'$ be a DIRICHLET region, and $G'_n + B'_n$ a sequence approximating G' in the sense of 5.6. Denote by F a function which is continuous on $G'+B'$, by h the solution of the DIRICHLET problem for $G'+B'$ with the boundary condition $h = F$ on B', and by h_n the solution of the DIRICHLET problem for G'_n with the boundary condition $h_n = F$ on B'_n. Then h_n approximates h in the following sense. Given $\varepsilon > 0$, we have an $n_0 = n_0(\varepsilon)$ such that $|h - h_n| < \varepsilon$ in $G'_n + B'_n$ for $n > n_0$. This follows by simple ε-arguments from the maximum-minimum principle for harmonic functions.

5.24. *Remarks on the formula of* GREEN. Let g be continuous together with its derivatives of the first and second order in a domain G. Consider a region $G'+B'$ comprised in G, such that B' consists of a finite number of non-intersecting smooth JORDAN curves. Take a point (x_0, y_0) in G', and take r small enough so that the closed circular disc with centre (x_0, y_0) and radius r is comprised in G'. Put (see 4.19 to 4.21 for notations)

$$l(x,y) = l(x,y;\, x_0, y_0), \qquad l_r(x,y) = l_r(x,y;\, x_0, y_0),$$

$$g^{(r)}(x_0, y_0) = \frac{1}{2\pi}\int_0^{2\pi} g(x_0 + r\cos\varphi,\ y_0 + r\sin\varphi)\, d\varphi.$$

Denote by H the solution of the DIRICHLET problem for G' with the boundary condition $H = l$ on B', and by h the solution of the DIRICHLET problem for G' with the boundary condition $h = g$ on B'. The functions H, l, l_r depend also upon (x_0, y_0), but this point will be kept fixed and therefore it is unnecessary to use notations like $H(x, y; x_0, y_0)$. The function h is independent of (x_0, y_0), and H, h, l are all independent of r. As g and B' are smooth, it is easy to justify the application of GREEN's identity (KELLOGG [1], p. 215) in deriving the formula

$$g^{(r)}(x_0, y_0) = -\frac{1}{2\pi}\iint\limits_{G'} [l_r(x, y) - H(x, y)]\, \Delta g(x, y)\, dx\, dy + h(x_0, y_0).$$

5.25. For $r \to 0$ we obtain the classical formula

$$g(x_0, y_0) = -\frac{1}{2\pi}\iint\limits_{G'} \mathfrak{G}(x, y; x_0, y_0)\, \Delta g(x, y)\, dx\, dy + h(x_0, y_0),$$

where $\mathfrak{G} = l - H$ is GREEN's function for G' with pole at (x_0, y_0). Conversely, an integration leads back to the formula of 5.24 which is more convenient in some applications.

5.26. Drop now the assumption that the boundary B' of G' is smooth and suppose only that $G' + B'$ is a DIRICHLET region. Otherwise, let all assumptions and notations stand as in 5.24. *Then the formula of 5.24 still holds.* This is easily proved, on the basis of 5.23, by applying the formula to a sequence of regions which approximate G' in the sense of 5.6.

5.27. If the function g of 5.26 is subharmonic in G, then the function h is the B. H. M. of g in G' (observe that g is continuous by assumption and use 5.4).

5.28. We proceed to discuss the question raised in 5.8. We start with the following theorem. Let u be subharmonic in a domain G. Denote by $G' + B'$ a DIRICHLET region comprised in G, and by $\bar{h}$ the B. H. M. of u in G'. Suppose that u is *harmonic* in G'. Then $\bar{h} = u$ in G' (RADÓ [4]).

5.29. To prove this theorem, consider the approximating functions $u_k^{(3)}$ defined in 2.21. If $G'' + B''$ is a region such that $G' + B' \subset G''$, $G'' + B'' \subset G$, then for large k the function $u_k^{(3)}$ is defined in G'' and is continuous there together with its derivatives of the first and second order. Also, $u_k^{(3)}$ is subharmonic, and hence $\Delta u_k^{(3)} \geqq 0$. By 2.24 we have

$$0 \leqq \iint\limits_{G'} \Delta u_k^{(3)}(x, y)\, dx\, dy < N$$

where N is a finite constant. If S is any closed set in G', then we have $u_k^{(3)} = u$ and $\Delta u_k^{(3)} = 0$ on S for large k (see 2.23). Take now any point (x_0, y_0) in G' and a small r. For large k, the function $u_k^{(3)}$

is then harmonic on the closed circular disc with centre (x_0, y_0) and radius r (see 2.23) and hence we have

$$\frac{1}{2\pi}\int_0^{2\pi} u_k^{(3)}(x_0 + r\cos\varphi, y_0 + r\sin\varphi)\,d\varphi = u_k^{(3)}(x_0, y_0) = u(x_0, y_0).$$

5.30. Using the preceding facts, we obtain from 5.26 the formula

$$u_k^{(3)}(x_0, y_0) = -\frac{1}{2\pi}\iint_{G'}[l_r(x, y) - H(x, y)]\,\Delta u_k^{(3)}(x, y)\,dx\,dy + h_k^{(3)}(x_0, y_0),$$

where $h_k^{(3)}$ is the solution of the DIRICHLET problem for G' with the boundary condition $h_k^{(3)} = u_k^{(3)}$ on B'.

5.31. As $u_k^{(3)}$ is continuous and $u_k^{(3)} \searrow u$ on B' (see 2.21), we have $h_k^{(3)} \searrow \bar{h}$ for $k \to \infty$, where $\bar{h}$ is the B.H.M. of u in G' (see 5.4).

5.32. Give now an $\varepsilon > 0$. Observe that $l_r - H$ is continuous on $G' + B'$ and $l_r - H = l - H = 0$ on B'. Hence we have a $\delta > 0$ such that $|l_r - H| < \varepsilon$ in $G' - S_\delta$, where S_δ denotes the (closed) set of all those points in G' whose distance from B' is $\geqq \delta$. We write now

$$\iint_{G'}[l_r(x, y) - H(x, y)]\,\Delta u_k^{(3)}(x, y)\,dx\,dy = \iint_{S_\delta} + \iint_{G'-S_\delta} = I_k^{(1)} + I_k^{(2)}.$$

By 5.29 we have then $I_k^{(1)} = 0$ and $|I_k^{(2)}| < \varepsilon N$ for large k. As ε is arbitrary, it follows that the integral in the formula of 5.30 converges to zero. The term $h_k^{(3)}(x_0, y_0)$ in that formula converges to $\bar{h}(x_0, y_0)$ (see 5.31). Thus the theorem of 5.28 follows from the formula of 5.30 for $k \to \infty$.

5.33. Denote by $G' + B'$ a region comprised in the domain G in which u is subharmonic. Consider a function h' which is harmonic in G' and define in G a function u' as follows: $u' = u$ in $G - G'$ and $u' = h$ in G'. If u' is subharmonic in G, then let us say that h' is *admissible* for u in G'. We have then the theorem: if $G' + B'$ is a DIRICHLET region comprised in G, then there exists in G' *exactly one* harmonic function which is admissible for u in G' (RADÓ [**4**]). The fact that there exists *at most one* admissible harmonic function follows immediately from 5.28 and 5.4. The fact that there exists *at least one* follows from the next theorem.

5.34. If $G' + B'$ is a DIRICHLET region comprised in G, then the best harmonic majorant $\bar{h}$ of u in G' (see 5.4) is admissible for u in G' (EVANS [**4**], part I, p. 237). This follows immediately from the definition of the best harmonic majorant.

5.35. If $G' + B'$ is a DIRICHLET region comprised in G and if $\bar{h}$ and h^* denote the B.H.M. and the L.H.M. of u in G', then $\bar{h} \equiv h^*$ (RADÓ [**4**]). On account of 5.33 and 5.34 this will be proved if we show that h^* is admissible for u in G', and this fact follows immediately from 5.34 and from the relation $u \leqq h^* \leqq \bar{h}$.

Chapter VI.

Representation of subharmonic functions in terms of potentials.

6. 1. Every sufficiently smooth function v can be represented as a potential plus a harmonic function (KELLOGG [1], p. 219). We shall formulate this fact in a form suitable for our purposes. Let v be continuous in a domain G together with its derivatives of the first and second order. Take a region $G' + B'$ in G, such that B' consists of a finite number of non-intersecting smooth JORDAN curves. From GREENS identity we obtain the formula

$$L(v; P; r) = \frac{1}{2\pi} \iint_{G'} l_r(P, Q)\, \Delta v(Q)\, d a_Q + h(P), \quad P \text{ in } G', \ r \text{ small},$$

where

$$h(P) = \frac{1}{2\pi} \int_{B'} \left(u(Q) \frac{\partial l(P, Q)}{\partial n_e} - l(P, Q) \frac{\partial u(Q)}{\partial n_e} \right) ds$$

is harmonic in G' (see 1.5, 4.19, 4.21, 4.17 for notations).

6. 2. The harmonic function h of 6.1 depends only upon the values of v on B' and in the vicinity of B' (see the explicit formula in 6.1). In particular, h is independent of r. For $r \to 0$ we obtain (cf. 4.33)

$$v(P) = \int_{G'} l(P, Q)\, d\mu(e_Q) + h(P), \quad P \text{ in } G',$$

where μ is the mass-distribution with density $-(1/2\pi)\Delta v$.

6. 3. It is a fundamental result of F. RIESZ that every subharmonic function admits of a representation of this form, regardless of its possible lack of smoothness. F. RIESZ ([5], part II) gave two proofs for this theorem. We shall sketch a simplified version, due to G. C. EVANS ([4], part I, p. 237), of the second proof of RIESZ.

6. 4. *A selection theorem* (special case of RADON [1], p. 1337; see also RIESZ [5], part II, p. 351). Let there be given on a closed set S a sequence of positive mass-distributions $\mu_k(e)$ such that $\mu_k(S)$ is less than some finite constant independent of k. Then there exists a subsequence μ_{k_r} and a positive mass-distribution $\mu(e)$ on S, such that

$$\int_S f(Q)\, d\mu_{k_r}(e_Q) \to \int_S f(Q)\, d\mu(e_Q)$$

for every function $f(Q)$ which is continuous on S. The subsequence μ_{k_r} is then said to converge *weakly* to μ on S.

6. 5. Consider now a function $u(x, y) = u(P)$ which is subharmonic in a domain G. Take a region $G' + B'$ comprised in G. Take an auxiliary region $G'' + B''$ such that $G' + B' \subset G''$, $G'' + B'' \subset G$ and B''

consists of a finite number of smooth non-intersecting JORDAN curves Γ_i'', $i = 1, 2, \ldots, j$. Denote by D_i'' a narrow doubly connected DIRICHLET domain which contains Γ_i'' in its interior and by $\bar{h}_i$ the B. H. M. of u in D_i''. Define a function $\bar{u}$ in G as follows: $\bar{u} = u$ in $G - \Sigma D_i''$ and $\bar{u} = \bar{h}_i$ in D_i'', $i = 1, 2, \ldots, j$. Then $\bar{u}$ is subharmonic in G (see 5.34). Also, $\bar{u}$ is harmonic on B'' and in the vicinity of B'', and $\bar{u} = u$ in and near to $G' + B'$.

6.6. Consider now the functions $\bar{u}_k^{(3)}(x, y) = A_{1/k}(x, y; \bar{u})$ defined in 2.21. We have by 6.1 the formula

$$L(\bar{u}_k^{(3)}; P; r) = \frac{1}{2\pi} \iint\limits_{G''} l_r(P, Q)\, \Delta \bar{u}_k^{(3)}(Q)\, d a_Q + h_k(P),$$

$$P \text{ in } G'',\ k \text{ large},\ r \text{ small}.$$

6.7. By 2.23 we have $\bar{u}_k^{(3)} = \bar{u}$ on B'' and in the vicinity of B''. *Hence h_k is independent of k*, because h_k depends only upon the values of $\bar{u}_k^{(3)}$ on and near to B'' (see 6.2). We can write therefore h instead of h_k.

6.8. Define, for large k,

$$\bar{\mu}_k(e) = \frac{1}{2\pi} \iint\limits_{e} \Delta \bar{u}_k^{(3)}(Q)\, d a_Q, \quad e \subset G'' + B''.$$

We have then by GREEN's identity

$$\bar{\mu}_k(G'' + B'') = \frac{1}{2\pi} \int\limits_{B''} \frac{\partial \bar{u}_k^{(3)}}{\partial n_e}\, ds = \frac{1}{2\pi} \int\limits_{B''} \frac{\partial \bar{u}}{\partial n_e}\, ds.$$

Hence we can apply the selection theorem of 6.4 and we obtain on $G'' + B''$ a positive mass-distribution $\mu(e)$, such that a certain subsequence $\bar{\mu}_{k_\nu}$ converges weakly to μ on $G'' + B''$.

6.9. By 6.7 and 4.17 the formula of 6.6 can be written in the form

$$L(\bar{u}_k^{(3)}; P; r) = \int\limits_{G''+B''} l_r(P, Q)\, d\mu_k(e_Q) + h(P), \quad P \text{ in } G'',$$

since $\mu_k(B'')$ is clearly equal to zero. For $k = k_\nu$, $\nu \to \infty$ we obtain by 6.8 and 2.21

$$L(\bar{u}; P; r) = \int\limits_{G''+B''} l_r(P, Q)\, d\mu(e_Q) + h(P), \quad P \text{ in } G''.$$

For P in G' we have $\bar{u} = u$ by 6.5, and for $r \to \infty$ it follows by 2.7 and 4.23 that

$$u(P) = \int\limits_{G''+B''} l(P, Q)\, d\mu(e_Q) + h(P) = \int\limits_{G'} + \int\limits_{G''+B''-G'} + h(P)$$

for P in G'. The second integral on the right is a harmonic function of P in G' (see 4.25). Hence we have the following theorem.

6.10. If u is subharmonic in a domain G, and if G' is a domain completely interior to G, then there exists in G' a positive mass-distribution $\mu(e)$ such that

$$u(P) = -\int_{G'} \log \frac{1}{PQ}\, d\mu(e_Q) + H(P), \quad P \text{ in } G',$$

where H is harmonic in G' (RIESZ [5], part II).

6.11. We shall see now that the distribution $\mu(e)$ is *unique*. Put

$$v(P) = -\int_{G'} \log \frac{1}{PQ}\, d\mu(e_Q), \quad P \text{ in } G'.$$

Take in G' any two JORDAN curves C_1, C_2 such that C_1 is enclosed by C_2 and the interior of C_2 is comprised in G'. We have then (see 5.11) $F(C_1, C_2; u) = F(C_1, C_2; v) + F(C_1, C_2; H)$. But $F(C_1, C_2; H) = 0$, since H is harmonic in and on C_2. Hence $F(C_1, C_2; v)$ is univocally determined by u, and by 5.20 and 4.14 it follows that $\mu(e)$ is univocally determined on G'.

6.12. More generally, consider two domains G_1', G_2' completely interior to G, and denote by $\mu_1(e), \mu_2(e)$ the distributions which correspond to u_1, u_2 in the sense of 6.10 and 6.11. Then $\mu_1(e) = \mu_2(e)$ for every set e of class (B) which is comprised in $G_1' G_2'$. This follows by a reasoning similar to that in 6.11.

6.13. Let G be a domain and $\mathring{\mu}(e)$ a set-function which is defined only for sets e which are *completely interior* to G (that is, the limit points of e are also comprised in G; we only consider sets e which are measurable in the BOREL sense). If otherwise $\mathring{\mu}(e)$ possesses the properties required in 4.7, then $\mathring{\mu}(e)$ will be called a *generalized* positive mass-distribution on G. For such a distribution $\mathring{\mu}$ it might happen that the least upper bound of $\mathring{\mu}(e)$, for all sets e completely interior to G, is equal to $-\infty$.

6.14. Let E be a set [measurable (B)] completely interior to G. Considered on E, the $\mathring{\mu}$ of 6.13 is clearly a positive mass-distribution in the original sense of 4.7. Hence we can consider on E STIELTJES-RADON integrals in terms of $\mathring{\mu}$.

6.15. If u is subharmonic in a domain G, then there exists on G a univocally determined generalized positive mass-distribution $\mathring{\mu}(e)$ (see 6.13), such that for every domain G' completely interior to G we have

$$u(P) = -\int_{G'} \log \frac{1}{PQ}\, d\mathring{\mu}(e_Q) + h(P), \quad P \text{ in } G',$$

where h is harmonic in G' (RIESZ [5], part II; cf. 5.22). This follows immediately from 6.10 and 6.12.

6.16. Consider a DIRICHLET region $G' + B'$ and a point P in G'. Then GREEN's function for G' with pole at P is defined by $\mathfrak{G}(P, Q) = \log(1/PQ) - H(P, Q)$ where $H(P, Q)$ is the solution of the DIRICHLET problem for G' with the boundary condition $H(P, Q) = \log(1/PQ)$ on B'. Consider next a general bounded domain G. Approximate G by a sequence $G_n + B_n$ of DIRICHLET regions as explained in 5.6. If P is a point in G, then P will be in G_n for large n. GREEN's function $\mathfrak{G}(P, Q)$ for G with pole at P is then defined by $\mathfrak{G}(P, Q) = \lim \mathfrak{G}_n(P, Q)$. $\mathfrak{G}(P, Q)$ is a finite, positive and harmonic function of Q in G, except for $Q = P$, and we have $\mathfrak{G}(P, Q) = \log(1/PQ) - H(P, Q)$, where $H(P, Q)$ is a harmonic function of Q in G, even for $Q = P$ (see KELLOGG [1], Chapter IX for information concerning GREEN's function).

6.17. Let there be given in a bounded domain G a positive mass-distribution $\mathring{\mu}(e)$ in the generalized sense of 6.13. Consider

$$v_k(P) = -\int_{G_k} \mathfrak{G}(P, Q)\, d\mathring{\mu}(e_Q), \qquad P \text{ in } G_k,$$

where the sequence G_k approximates G in the sense of 5.6. We have more explicitly (see 6.16)

$$v_k(P) = -\int_{G_k} \log\frac{1}{PQ}\, d\mathring{\mu}(e_Q) + \int_{G_k} H(P, Q)\, d\mathring{\mu}(e_Q).$$

Thus v_k appears as the sum of a subharmonic function and of a harmonic function. Hence v_k is subharmonic in G_k. Clearly v_k decreases if k increases. By 3.6, either $v_k \to -\infty$ everywhere in G, or v_k converges to a subharmonic function v in G. In the latter case we write

$$v(P) = -\int_G \mathfrak{G}(P, Q)\, d\mathring{\mu}(e_Q).$$

On account of its definition, this integral is therefore a subharmonic function of P whenever it exists. The value of the integral is easily seen to be independent of the sequence G_k.

6.18. Consider now a function u which is subharmonic in the bounded domain G. Denote by $\mathring{\mu}$ the generalized distribution which corresponds to u in the sense of 6.15. Then the integral of 6.17 exists if and only if we have some harmonic function which is $\geqq u$ in G. If this condition is satisfied then

$$\int_G \mathfrak{G}(P, Q)\, d\mathring{\mu}(e_Q) \leqq h(P) - u(P)$$

for every harmonic function which is $\geqq u$ in G (RIESZ [5], part II). Proof. Suppose first that we have a harmonic function $h \geqq u$ in G. Take a domain G' completely interior to G and introduce again the auxiliary region $G'' + B''$ and the auxiliary functions $\bar{u}$, $\bar{u}_k^{(3)}$ as in 6.5 and 6.6. Denote by $\mathfrak{G}_r''(P, Q)$ the function obtained from GREEN's function for G'' if we replace $l(P, Q)$ by $l_r(P, Q)$ (see 4.21). We have then by 5.24, 6.7, 6.8 the formula

$$L(\bar{u}_k^{(3)}; P; r) = -\int_{G''+B''} \mathfrak{G}_r''(P, Q)\, d\mu_k(e_Q) + h_k(P), \qquad P \text{ in } G'',\ k \text{ large},\ r \text{ small},$$

where $\bar{h}_k$ is the solution of the DIRICHLET problem for G'' with the boundary condition $\bar{h}_k = \bar{u}_k^{(3)}$ on B''. For P in G' it follows, by a reasoning similar to that in 6.5 to 6.9, that

$$L(\bar{u}; P; r) = -\int\limits_{G''+B''} \mathfrak{G}_r''(P, Q)\, d\mu(e_Q) + \bar{h}(P), \qquad P \text{ in } G',$$

where the harmonic function $\bar{h}$ is determined by the condition $\bar{h} = \bar{u}$ on B''. It follows from the definition of $\bar{u}$ (see 6.5) that $\bar{h} \leqq h$ in $G'' + B''$ and $u \leqq \bar{u}$ in G. We have therefore

$$h(P) - L(u; P; r) \geqq \bar{h}(P) - L(\bar{u}; P; r) \geqq \int\limits_{G'} \mathfrak{G}_r''(P, Q)\, d\mathring{\mu}(e_Q),$$

since $\mathring{\mu}(e) = \mu(e)$ on G', by 6.8, 6.9, 6.10, 6.15. Denote by $\mathfrak{G}_r(P, Q)$ the function obtained from GREEN's function for G if we replace $l(P, Q)$ by $l_r(P, Q)$ (see 4.21). Let G'' approach G in the sense of 5.6. Then $\mathfrak{G}_r'' \nearrow \mathfrak{G}_r$ and by a well-known theorem oi DINI the convergence is uniform on every closed set in G (and hence on every set completely interior to G), since $\mathfrak{G}_r''$ and $\mathfrak{G}_r$ are continuous. We obtain for $G'' \to G$ the inequality $h(P) - L(u; P; r) \geqq \int\limits_{G'} \mathfrak{G}_r(P, Q)\, d\mathring{\mu}(e_Q)$. For $r \to 0$ it follows, by 6.17, that $h(P) - u(P) \geqq \int\limits_{G'} \mathfrak{G}(P, Q)\, d\mathring{\mu}(e_Q)$ for P in G'. As G' is any domain completely interior to G, the preceding inequality proves both the existence of the integral $\int\limits_{G} \mathfrak{G}(P, Q)\, d\mathring{\mu}(e_Q)$ and the inequality asserted in the theorem. Conversely, suppose that the preceding integral exists. Consider the functions v_k of 6.17 relative to the distribution $\mathring{\mu}$ which corresponds to the given subharmonic function u. By the definition of $\mathring{\mu}$ (see 6.15) we have $u(P) = v_k(P) + h_k(P)$ for P in G_k, where h_k is harmonic in G_k. If k increases, v_k decreases and hence h_k increases. By the theorem of HARNACK, $h_k \to h^*$ where either h^* is $\equiv +\infty$ in G or h^* is harmonic in G. Clearly the first case is incompatible with our present assumptions. For $k \to \infty$ we obtain therefore (cf. 6.17)

$$u(P) = -\int\limits_{G} \mathfrak{G}(P, Q)\, d\mathring{\mu}(e_Q) + h^*(P), \quad P \text{ in } G.$$

But $\mathring{\mu}$ and $\mathfrak{G}$ are both positive, and hence $h^* \geqq u$ in G. The existence of a harmonic majorant for u in G is proved.

6.19. The harmonic function h^* of the last formula of 6.18 is actually the least harmonic majorant of u in G. Indeed, if h is any harmonic majorant of u in G, then we have by 6.18

$$h^*(P) = u(P) + \int\limits_{G} \mathfrak{G}(P, Q)\, d\mathring{\mu}(e_Q) \leqq u(P) + (h(P) - u(P)) = h(P).$$

We have therefore the theorem: If u is subharmonic in a bounded domain G and if there exists a harmonic majorant for u in G, then u can be represented in the form

$$u(P) = -\int_G \mathfrak{G}(P, Q)\, d\mathring{\mu}(e_Q) + h^*(P), \quad P \text{ in } G,$$

where h^* is the least harmonic majorant of u in G and $\mathring{\mu}$ is a generalized positive mass-distribution on G (RIESZ [5], part II).

6.20. If u is a *smooth* subharmonic function, then the corresponding distribution can be expressed in terms of the LAPLACian Δu (see 6.2 and 6.11). It is then natural to expect that the preceding theorems can be discussed in terms of the generalized LAPLACians introduced by various authors. It seems that no explicit discussion was given as yet on this basis (cf. the remarks of F. RIESZ in WIENER [1], p. 7).

6.21. In the light of the theorem of 6.10, the theory of subharmonic functions appears as a chapter in potential theory. It is beyond the scope of this report to follow up the implications of this situation. The reader desiring further information will find a wealth of interesting material and a large number of references in FROSTMAN [1], EVANS [4], KELLOGG [1].

6.22. The theorem of 6.10 implies that the study of subharmonic functions *in the small* can be based upon a study of the potential $v(P) = \int \log PQ\, d\mu(e_Q)$. In the way of illustration, we mention two results obtained in this manner. According to EVANS ([4], part I, pp. 233—235) the potential $v(P)$ is an absolutely continuous function of x for almost every y and an absolutely continuous function of y for almost every x. As a consequence, the partial derivatives v_x and v_y exist almost everywhere. It follows by further discussion that v_x and v_y are summable on every bounded measurable set. The application to subharmonic functions is immediate on account of 6.10.

6.23. Using the notations of 6.15, consider the integral mean $A_r(x, y; u)$ (see 2.19). As $A_r(x, y; u)$ is again subharmonic, it will give rise to a distribution $\mathring{\mu}_r(e)$ in the sense of 6.15. It might be expected that $\mathring{\mu}_r(e)$ will be smoother than the distribution $\mathring{\mu}(e)$ corresponding to u itself. By means of 6.15 it follows from results of THOMPSON [1] that $\mathring{\mu}_r(e)$ is a distribution with a summable density $\delta_r(P)$ given by

$$\delta_r(P) = \frac{1}{r^2\pi}\, \mathring{\mu}(C(r; P)),$$

where $C(r; P)$ denotes the interior of the circle with centre P and radius r. The proof depends upon a discussion of changes in the order of integrations in iterated STIELTJES-RADON integrals.

Chapter VII.

Analogies between harmonic and subharmonic functions.

7.1. The general theory of subharmonic functions, as sketched in the preceding Chapters, was based largely upon a few elementary properties of harmonic functions. Practically every paper quoted in this report contains interesting developments concerned with the implications of more involved properties of harmonic functions. The purpose of this Chapter is to give a picture of the results obtained in this direction. The reader will note that the proofs sketched in the sequel do not always apply in the case of more than two variables. Such situations lead to interesting problems, some of which seem to be quite difficult. As a first topic, we shall consider *isolated singularities of subharmonic functions*. If u is known to be subharmonic in the vicinity of a point (x_0, y_0), this point itself being excluded, then (x_0, y_0) will be called an isolated singular point of u. Without loss of generality we can assume that (x_0, y_0) is the point $O = (0, 0)$. We shall review presently some results of BRELOT [1]. Various details of the following presentation are based on unpublished remarks of S. SAKS.

7.2. (See 1.5 for notations.) Put $L(u; r) = L(u; 0, 0; r)$, $\lambda = u/\log(1/r)$, $L(\lambda; r) = L(u; r)/\log(1/r)$. By 2.5, $L(u; r)$ is a convex function of $\log r$ and hence of $\log(1/r)$ for small r. Using some simple properties of convex functions, we obtain a number of facts concerning $L(u; r)$ and $L(\lambda; r)$ (BRELOT [1], pp. 23—37), some of which we shall list now explicitly.

7.3. For $r \to 0$ both $L(u; r)$ and $L(\lambda; r)$ converge to definite (not necessarily finite) limits which will be denoted by $L(u; 0)$ and $L(\lambda; 0)$ respectively. For small values of r both $L(u; r)$ and $L(\lambda; r)$ are *monotonic*, and for $r \searrow 0$ either $L(u; r) \nearrow L(u; 0) = +\infty$ or $L(u; r) \searrow L(u; 0) \geqq -\infty$, and either $L(\lambda; r) \nearrow L(\lambda; 0) \leqq +\infty$ or $L(\lambda; r) \searrow L(\lambda; 0) > -\infty$. Note that if $L(u; r)$ increases for $r \searrow 0$ then always $L(u; 0) = +\infty$, and if $L(\lambda; r)$ decreases for $r \searrow 0$ then always $L(\lambda; 0) > -\infty$.

7.4. We shall use $\overset{+}{a}$ to denote the greater one of the numbers a and zero. Clearly $a \leqq \overset{+}{a} \leqq |a|$ and $|a| = 2\overset{+}{a} - a$.

7.5. Let us recall a few facts concerning isolated singularities of *harmonic* functions. If $h(P)$ is harmonic in the vicinity of O with the possible exception of O itself, then we have the expansion (see KELLOGG [1], Chapter XII)

$$h(P) = h_0(P) + \gamma \log \frac{1}{OP} + \sum_{n=1}^{\infty} \frac{\alpha_n \cos n\varphi + \beta_n \sin n\varphi}{OP^n} = h_0(P) + h_1(P),$$

where $h_0(P)$ is harmonic even at O, $h_1(P)$ is harmonic in the whole plane with the possible exception of O, φ is the polar angle determined by $x = OP\cos\varphi$, $y = OP\sin\varphi$, and γ, α_n, β_n are constants. Suppose that $h \geqq 0$ in the vicinity of O. Then $\alpha_n = \beta_n = 0$ for $n = 1, 2, \ldots$ (see BRELOT [1] for references; see also RIESZ [5], part II, p. 350). Indeed, we have $L(h\cos n\varphi; r) = h_0(O) + \gamma\log(1/r) + \alpha_n/(2r^n)$. Hence $\alpha_n = 2\lim r^n L(h\cos n\varphi; r)$, $r \to 0$. But $|L(h\cos n\varphi; r)| \leqq L(|h|; r) = L(h; r) = h_0(O) + \gamma\log(1/r)$. Thus $r^n L(h\cos n\varphi; r) \to 0$ for $r \to 0$, and hence $\alpha_n = 0$. The coefficient β_n is discussed in the same way.

7.6. Until further notice, u denotes a subharmonic function which has an isolated singularity at O. The following remark will be useful in the sequel. Consider, for small r'', a circular ring $R: 0 < r' < (x^2+y^2)^{1/2} < r''$, and denote by $\bar{h}$ the B.H.M. of u in R (see 5.4). Then (cf. 1.12, 1.13) $L(\bar{h}; r) = L(\bar{h}; 0, 0; r)$ is a linear function $a\log(1/r) + b$ of $\log(1/r)$ and we have $a\log(1/r') + b = L(u; r')$, $a\log(1/r'') + b = L(u; r'')$. Suppose we are given inequalities $L(u; r') \leqq A\log(1/r') + B$, $L(u; r'') \leqq A\log(1/r'') + B$, where A, B are constants. Clearly, it follows that $L(\bar{h}; r) \leqq A\log(1/r) + B$ for $r' < r < r''$.

7.7. Since u is subharmonic in the vicinity of O, the function $\overset{+}{u} = \overline{u, 0}$ is also subharmonic there (see 3.4). We shall use the symbols $L(\overset{+}{u}; r)$, $L(\overset{+}{\lambda}; r)$ in the sense of 7.2.

7.8. We shall use $V - O$ to denote a vicinity of O, less O itself. It is assumed that u is subharmonic in $V - O$ and on the boundary of $V - O$, except for the point O. Vicinities of the form $0 < x^2 + y^2 < \varrho^2$ will be denoted by $V_\varrho - O$. It will be understood that ϱ is so small that $L(u; r)$, $L(\lambda; r)$, $L(\overset{+}{u}; r)$, $L(\overset{+}{\lambda}; r)$ are *monotonic* for $0 < r < \varrho$ (see 7.3 and 7.7).

7.9. Suppose that in a vicinity $V_\varrho - O$ we have a harmonic majorant H for u, and let $H = H_0 + H_1$ be the expansion of H (cf. 7.5). Let $V - O$ be any vicinity which contains $V_\varrho - O$. Then u has in $V - O$ a harmonic majorant of the form $H_1 + \text{const}$. This follows immediately from the fact that u is upper semi-continuous and H_1 is continuous in $V - O$ and on the boundary of $V - O$, the point O being excluded.

7.10. Generally there will not exist a harmonic majorant for u in the vicinity of O. However, if we have a harmonic majorant $H^{(1)}$ in a vicinity $V_1 - O$, then we also have a harmonic majorant in any other vicinity $V_2 - O$ (BRELOT [1], p. 32). To see this, take a vicinity $V_\varrho - O$ comprised both in $V_1 - O$ and in $V_2 - O$. Observe that $H^{(1)}$ is also a majorant in $V_\varrho - O$ and apply 7.9.

7.11. Suppose u has a harmonic majorant in a vicinity $V_\varrho - O$. Then (see 5.7) we have in $V_\varrho - O$ a L.H.M. h^*. Let $h^* = h_0 + h_1$ be the expansion of h^* (see 7.5). Consider now any vicinity $\tilde{V} - O$

containing $V_\varrho - O$. By 7.9 we have in $\tilde{V} - O$ a harmonic majorant $\tilde{H}$ of the form $h_1 +$ const., and hence the L. H. M. $\tilde{h}^*$ of u in $\tilde{V} - O$ satisfies an inequality $\tilde{h}^* < h_1 +$ const. in the vicinity of O. We have therefore $0 \leqq h^* - \tilde{h}^* \leqq h_0 +$ const. in the vicinity of O. Consequently (cf. 7.5) $h^* - \tilde{h}^*$ is harmonic even at O.

7.12. If h_1^*, h_2^* are the L. H. M. of u in the vicinities $V_1 - O$, $V_2 - O$ respectively, then $h_1^* - h_2^*$ is harmonic even at O (Brelot [1], p. 32). Proof. Take a vicinity $V_\varrho - O$ comprised both in $V_1 - O$ and in $V_2 - O$ and apply 7.11.

7.13. Suppose that u has a harmonic majorant in a vicinity of O. By 7.6 we have then a L. H. M. for u in every vicinity $V - O$ and by 7.11, 7.12, 7.5 the constants γ, α_n, β_n have the same values in the expansions of all these least harmonic majorants.

7.14. u has a harmonic majorant in the vicinity of O if and only if $L(\lambda; 0)$ (see 7.3) is finite. If this condition is satisfied, then the constant γ of 7.13 is equal to $L(\lambda; 0)$ (Brelot [1], p. 32). Proof. The necessity of the condition is obvious. To prove the sufficiency, assume that $L(\lambda; 0)$ is finite. Give an $\varepsilon > 0$ and take a small r_0 such that $L(\lambda; r) < L(\lambda; 0) + \varepsilon$ for $0 < r < r_0$. Take any r such that $0 < r < r_0$ and take a sequence r_n such that $r > r_1 > r_2 > \cdots \to 0$. Denote by $\bar{h}_n$ the B. H. M. of u in the ring $r_n < (x^2 + y^2)^{1/2} < r_0$. We have then $L(u; r_n) < (L(\lambda; 0) + \varepsilon) \log(1/r_n)$, $L(u; r_0) < (L(\lambda; 0) + \varepsilon) \log(1/r_0)$, and hence, by 7.6, $L(\bar{h}^n; r) < (L(\lambda; 0) + \varepsilon) \log(1/r)$. This shows that $\bar{h}_n$ cannot converge to $+\infty$ everywhere. By 5.5 we have $\bar{h}_{n+1} \geqq \bar{h}_n \geqq u$ in the ring $r_n < (x^2 + y^2)^{1/2} < r_0$. Hence, by the theorem of Harnack, $\bar{h}_n$ converges in $V_{r_0} - O$ to a harmonic function $h^* \geqq u$, and the existence of a harmonic majorant is proved. By 5.7, h^* is the L. H. M. of u in $V_{r_0} - O$. Also, the inequality $L(\bar{h}_n; r) < (L(\lambda; 0) + \varepsilon) \log(1/r)$ implies that we have $L(h^*; r) \leqq (L(\lambda; 0) + \varepsilon) \log(1/r)$ for $0 < r < r_0$. To estimate the constant γ (cf. 7.5, 7.13) in the expansion of h^*, observe that

$$\gamma = \lim_{r \to 0} \frac{L(h^*; r)}{\log \frac{1}{r}} \leqq L(\lambda; 0) + \varepsilon .$$

On the other hand, $u \leqq h^*$ in $V_{r_0} - O$ and hence

$$L(\lambda; 0) = \lim_{r \to 0} \frac{L(u; r)}{\log \frac{1}{r}} \leqq \lim_{r \to 0} \frac{L(h^*; r)}{\log \frac{1}{r}} = \gamma .$$

As ε is arbitrary, it follows that $\gamma = L(\lambda; 0)$.

7.15. If u is bounded from above in the vicinity of O, then u is subharmonic even at O (Brelot [1], p. 27). Instead of reproducing the proof of Brelot, let us observe that this follows-immediately from the theorems of 3.35 and 3.37 on almost subharmonic functions. Indeed, consider for $n = 1, 2, \ldots$ the function $u_n(P) = u(P) - (1/n)\log(1/OP)$ for $P \neq O$, $u_n(O) = -\infty$. Clearly, u_n is subharmonic in a small disc $D: x^2 + y^2 < r^2$, even at O, since by assumption $u < M$ in $D - O$, where M is some finite constant. We have $u_n \leqq u_{n+1} < M$ in D, since $u_{n+1}(O) = -\infty$. Hence, by 3.37, the limit function $u^* = \lim u_n$ is almost subharmonic in D. We have therefore in D a subharmonic function $\bar{u}$ such that $\bar{u} = u^*$ almost everywhere in D. But $u^* = u$ in $D - O$, and hence $\bar{u} = u$ in $D - O$ by 3.35. As $\bar{u}$ is subharmonic even at O, the theorem is proved.

7.16. If u has a harmonic majorant H in the vicinity of O, then $v = u - H$ is subharmonic even at O (Brelot [1], p. 35). As $v \leqq 0$ in the vicinity of O, this follows immediately from 7.15.

7.17. (See 7.2, 7.3, 7.7 for notations). If $L(\overset{+}{u}; 0)$ is finite, then u remains subharmonic at O (Brelot [1], pp. 34—35). More generally, if $L(\overset{+}{\lambda}; 0) = 0$, then u remains subharmonic at O (Saks, unpublished). Proof. By 7.7 and 7.14, $\overset{+}{u}$ has a harmonic majorant in the vicinity of O. Denote by H^* the least harmonic majorant of $\overset{+}{u}$ in a vicinity $V_\varrho - O$. Then $H^* \geqq \overset{+}{u} \geqq 0$ and hence by 7.5 we have $H^*(P) = H_0(P) + \Gamma\log(1/OP)$, where H_0 is harmonic even at O. By 7.14 we have $\Gamma = L(\overset{+}{\lambda}; 0)$, which is equal to zero by assumption. Hence H^* is harmonic even at O. Consequently $\overset{+}{u}$ is bounded from above in the vicinity of O. As $u \leqq \overset{+}{u}$, the theorem follows now by 7.15.

7.18. The work of Brelot contains a number of further results and applications which cannot be reproduced here. We shall review presently certain results concerned with generalizations of properties of harmonic functions in the vicinity of the boundary of the domain of definition. The first results in this direction were obtained by Littlewood [**2, 3, 4**] and Evans [**2**]. These results were later on extended by Evans [**3**] and Privaloff [**1, 2**]. Evans obtained his results by methods in Potential Theory. Privaloff extended some results obtained by Littlewood in the special case of the circle to more general regions. In the way of illustration we shall give a few details concerning the work of Littlewood.

7.19. We shall refer in the sequel to the inequality of Hölder: if f and g are non-negative functions, and if $p > 0$, $q > 0$ are exponents such that $(1/p) + (1/q) = 1$, then $\int fg \leqq (\int f^p)^{1/p} (\int g^q)^{1/q}$, whenever the integrals involved exist in the Lebesgue sense (for a particularly elegant proof, see Riesz [**6**]).

7.20. Suppose that u is subharmonic for $x^2 + y^2 < 1$ and that $L(|u|^p; 0, 0; \varrho) < G^p$ for $\varrho < 1$, where G is a constant and $p > 1$. If u were *harmonic*, then these assumptions would imply the existence of a function $w(\Theta)$ such that $u(\varrho\cos\Theta, \varrho\sin\Theta) \to w(\Theta)$ for $\varrho \to 1$ and almost every Θ, and

$$\int_0^{2\pi} |u(\varrho\cos\Theta, \varrho\sin\Theta) - w(\Theta)|^p \, d\Theta \to 0 \quad \text{for} \quad \varrho \to 1$$

(RIESZ [2]). In the case of a general subharmonic function LITTLEWOOD obtained the following results.

7. 21. Under the assumptions of 7. 20 there exists a function $w(\Theta)$ such that

$$\int_0^{2\pi} |u(\varrho\cos\Theta, \varrho\sin\Theta) - w(\Theta)|^q \, d\Theta \to 0, \quad \varrho \to 1,$$

for every exponent $0 < q < p$ (LITTLEWOOD [2]). Proof. On account of the inequality of HÖLDER it is sufficient to consider the case $1 < q < p$. Denote by $\bar{h}_r$ the B. H. M. of u for $x^2 + y^2 < r^2$. It follows by the inequality of HÖLDER from the formula of POISSON for $\bar{h}_r$ (cf. 5. 4) that $\bar{h}_r$ satisfies an inequality of the same form as u. It follows that for $r \to 1$ the sequence $\bar{h}_r$ cannot converge to $+\infty$ everywhere. Hence (see 5.7 and 1. 4) there exists for u a L. H. M. h^* in $x^2 + y^2 < 1$ which satisfies an inequality of the same form as u. By the theorem of F. RIESZ quoted in 7. 20 we have therefore a function $w(\Theta)$ such that

$$\int_0^{2\pi} |h^*(\varrho\cos\Theta, \varrho\sin\Theta) - w(\Theta)|^p \, d\Theta \to 0$$

for $\varrho \to 1$. As $u \leqq \bar{h}_r \nearrow h^*$ for $r \nearrow 1$, it follows from the preceding facts, by repeated application of the inequality of HÖLDER, that $w(\Theta)$ satisfies the theorem.

7. 22. If u is subharmonic in $x^2 + y^2 < 1$ and $L(|u|; 0, 0; \varrho) < M$ for $\varrho < 1$ (M a finite constant), then $\lim u(\varrho\cos\Theta, \varrho\sin\Theta)$, $\varrho \to 1$, exists and is finite for almost every Θ (LITTLEWOOD [3]). This theorem is related to the theorems in Chapter VI as follows. Establish first the existence of the L. H. M. h^* of u in $x^2 + y^2 < 1$ as in 5. 21. Observe next that the assumption concerning u implies that $L(\overset{+}{u}; 0, 0; \varrho)$ is also bounded for $\varrho < 0$. Hence, for the same reasons as in the case of u, there exists a L. H. M. H^* for $\overset{+}{u}$ in $x^2 + y^2 < 1$, and we have $H^* \geqq \overset{+}{u} \geqq 0$, $H^* \geqq h^*$ in $x^2 + y^2 < 1$. By 6. 20 we have for u the representation

$$u(P) = -\int_{x^2+y^2<1} \mathfrak{G}(P, Q) \, d\mathring{\mu}(e_Q) + h^*(P) = v(P) + h^*(P)$$

for $x^2 + y^2 < 1$, where $\mathfrak{G}$ is GREEN's function for the unit circle. We can write $h^* = H^* - (H^* - h^*)$. Thus h^* appears as the difference of two *non-negative* harmonic functions, and hence h^* has a definite

finite radial limit along almost every radius (see, for references covering also the case of more than two variables, GARRETT [**1**]). Thus the problem is reduced to the discussion of $v(P)$. LITTLEWOOD shows that $v(P)$ has a radial limit equal to zero for almost every radius. The proof depends upon a number of inequalities concerning GREEN's function of the unit circle.

7.23. LITTLEWOOD constructed explicit examples which show that 1) in the theorem of 7.21 the condition $q < p$ cannot be replaced by $q \leqq p$, 2) the theorem of 7.21 is not valid for $q = p = 1$, and 3) for $0 < p < 1$ there does not exist, generally, a radial limit $w(\Theta)$, either in the sense of convergence almost everywhere or in the sense of convergence in the mean with respect to some exponent (LITTLEWOOD [**4**]).

7.24. Theorems on harmonic functions may involve *pairs of conjugate harmonic functions*, that is analytic functions of a complex variable. It is not clear a priori that such theorems can be extended to subharmonic functions of two or more variables. Questions of this type were discussed in considerable detail by PRIVALOFF, who generalized a number of theorems concerned with analytic functions of a complex variable (PRIVALOFF [**3, 4**]). In the way of illustration, we quote two of his theorems for the case of three independent variables. *Theorem.* Let u be subharmonic in a domain G in three-dimensional EUCLIDean space. Suppose that the boundary B of G can be divided into two parts B_1, B_2 in such a way that $\overline{\lim}\, u(P) \leqq M_k$ if P approaches any point of B_k, $k = 1, 2$. Let $G' + B'$ be a region comprised in G. Then there exist two constants s and t, $0 < s < 1$, $0 < t < 1$, depending only upon G and $G' + B'$, such that in $G' + B'$ we have $u \leqq t M_1 + (1 - t) M_2$ if $M_1 \leqq M_2$ and $u \leqq s M_1 + (1 - s) M_2$ if $M_1 \geqq M_2$. *Theorem.* Let u be subharmonic in a domain G in three-dimensional EUCLIDean space. Suppose that there exists a finite constant M such that $\overline{\lim}\, u(P) \leqq M$ if P approaches any point on the boundary of G, with the possible exception of a denumerable set of boundary points $Q_n, n = 1, 2, \ldots$ At these exceptional points it is known that $u(P) - \sigma/(PQ_n) \to -\infty$ for *every* $\sigma > 0$ if P approaches Q_n. Then $u \leqq M$ in G (this wording, due to SAKS, is somewhat more general than the original wording of PRIVALOFF). Let us sketch a simple proof (due to SAKS) of the second theorem. Consider in G the function $u_\varepsilon(P) = u(P) - \sum_{n=1}^{\infty} \varepsilon/(2^n PQ_n)$, $\varepsilon > 0$. Clearly, the infinite series converges in G, the convergence being uniform in every region $G' + B' \subset G$. Hence, by 3.3 and 2.15, u_ε is subharmonic in G. Clearly, $\overline{\lim}\, u_\varepsilon(P) \leqq M$ if P approaches *any* boundary point of G. By 1.15 we have therefore $u_\varepsilon(P) \leqq M$ in G. For P fixed and $\varepsilon \to 0$ it follows that $u \leqq M$ in G.

7. 25. To illustrate results of a different type, we quote the following theorem. Let Γ be a circle and C a convex curve strictly interior to Γ. Suppose that u is positive, continuous and subharmonic in and on Γ. Then

$$\int_C u^\lambda ds \leq 4 \int_\Gamma u^\lambda ds \text{ for } \lambda \geq 2, \quad \text{and} \quad \int_C u^\lambda ds \leq \frac{A}{\lambda - 1} \int_\Gamma u^\lambda ds \text{ for } 1 < \lambda < 2,$$

where A is an absolute constant (FRAZER [**1**]; this is a generalization of previous results of GABRIEL [**1, 2**], who generalized earlier results of FEJÉR-RIESZ [**1**]. See also RIESZ [**8**]). Proof. We can clearly assume that Γ is the unit circle $x^2 + y^2 = 1$. Suppose first that $\lambda = 2$. Denote by $\bar{h}$ the B. H. M. of u in Γ. In Γ we have an analytic function $f(z)$ whose real part is equal to $\bar{h}$, say $f(z) = \bar{h} + i h$, $z = x + i y$. We can suppose that $h(O) = \bar{h}(O)$. If Γ_r is a concentric circle with radius r, such that r is slightly less than 1, then we have by a theorem of GABRIEL [**1**]

$$\int_C |f|^2 ds \leq 2 \int_{\Gamma_r} |f|^2 ds.$$

We have $\Re f(0)^2 = 0$, since $h(O) = \bar{h}(O)$. Using Γ_r to express $f(0)^2$ by the formula of CAUCHY, we obtain

$$\int_{\Gamma_r} \bar{h}^2 ds = \int_{\Gamma_r} h^2 ds.$$

We can write now

$$\int_C u^2 ds \leq \int_C \bar{h}^2 ds \leq \int_C |f|^2 ds \leq 2 \int_{\Gamma_r} |f|^2 ds = 4 \int_{\Gamma_r} \bar{h}^2 ds \xrightarrow[r \to 1]{} 4 \int_\Gamma \bar{h}^2 ds = 4 \int_\Gamma u^2 ds,$$

and the theorem is proved for the special case $\lambda = 2$. If $\lambda > 2$, then apply the preceding result to $u^{\lambda/2}$ which is subharmonic by 3. 23. The case $1 < \lambda < 2$ is discussed in a similar fashion. For further theorems of this type see FRAZER [**2, 3, 4**].

7. 26. Clearly, the method used in 7. 25 does not apply in the case of more than two variables. To illustrate a somewhat different situation, we consider a result obtained by SAKS [**1**] as a corollary to more general theorems. Denote by G a *simply connected* domain in the interior K of the unit circle $x^2 + y^2 = 1$. Let u be subharmonic in G and suppose that $u(P) \to -\infty$ if u approaches any boundary point of G which is *interior* to K. Then $G \equiv K$. This theorem is closely related to recent results of EVANS [**5**]. The method of EVANS suggests the following proof. Define a function $\bar{u}$ as follows: $\bar{u} = u$ in G and $\bar{u} = -\infty$ in $K - G$. Then u is subharmonic in K (see 1. 1 and 2. 3). Hence the set $K - G$ cannot have *interior* points (see 1. 8). If $K \not\equiv G$, then we can assume that the centre of K is not in G. Apply now the transformation $w = \sqrt{z}$, $z = x + i y$ (cf. RADÓ [**1**], lemma on p. 2). As G is *simply connected,* we have a single-valued analytic branch of

$\sqrt{z}$ in G, and the transformation leads to a domain G' and a subharmonic function u', such that the assumptions of the theorem are satisfied by G' and u', and such that $K - G'$ *does* have interior points. This contradicts a preceding conclusion, and the theorem is proved.

SAKS goes on to show that the preceding theorem is *not* valid in three-dimensional space. Example: consider

$$u(x, y, z) = -\int_0^1 \frac{ds}{r}, \quad r = [(x-s)^2 + y^2 + z^2]^{1/2},$$

in the domain G consisting of all points (x, y, z) in $x^2 + y^2 + z^2 < 1$, except the points $0 \leqq x < 1$, $y = 0$, $z = 0$. We have here one of the many instances where the existence of the transformations $w = \sqrt[n]{z}$ leads to theorems in the plane which cannot be extended to spaces of higher dimensions.

References.

The numbers in round brackets denote chapters and paragraphs.

BECKENBACH, E. F.: [1] On subharmonic functions, Duke Math. J., vol. 1, 1935, pp. 481—483 (*2.13*, *4.5*).

BECKENBACH, E. F., and RADÓ, T.: [1] Subharmonic functions and minimal surfaces, Trans. Amer. Math. Soc., vol. 35, 1933, pp. 648—661 (*4.4*).

[2] Subharmonic functions and surfaces of negative curvature, Trans. Amer. Math. Soc., vol. 35, 1933, pp. 662—664 (*3.25*, *3.26*, *4.5*).

BLASCHKE, W.: [1] Ein Mittelwertsatz und eine kennzeichnende Eigenschaft des logarithmischen Potentials, Ber. Verh. Sächs. Akad. Wiss., Leipziger, vol. 68, 1916, pp. 3—7 (*3.7*).

BOREL, E.: [1] Leçons sur les fonctions de variables réelles, Paris, 1905 (*4.6*).

BRELOT, M.: [1] Étude des fonctions sousharmoniques au voisinage d'un point, Actualités scientifiques et industrielles, vol. 139, 1934 (*1.14*, *3.13*, *4.1*, *4.34*, *5.8*, *7.1*, *7.2*, *7.5*, *7.10*, *7.12*, *7.14—7.17*).

CARLEMAN, T.: [1] Zur Theorie der Minimalflächen, Math. Z., vol. 9, 1921, pp. 154 to 160 (*3.26*).

DANIELL, P. J.: [1] A general form of integral, Annals of Math., vol. 19, 1917, pp. 279—294 (*4.23*).

EVANS, G. C.: [1] Fundamental points of potential theory, Rice Institute Pamphlets, vol. 7, 1920, pp. 252—329 (*5.16*).

[2] Discontinuous boundary value problems of the first kind for Poisson's equation, Amer. J. Math., vol. 51, 1929, pp. 1—18 (*7.18*).

[3] Complements of potential theory, part II, Amer. J. Math., vol. 55, 1933, pp. 29—49 (*7.18*).

[4] On potentials of positive mass, part I and II, Trans. Amer. Math. Soc., vol. 37, 1935, pp. 226—253 and vol. 38, 1935, pp. 201—236 (*1.2*, *4.6*, *5.22*, *5.34*, *6.3*, *6.21*, *6.22*).

[5] Potentials and positively infinite singularities of harmonic functions, Mh. Math. Physik, vol. 43, 1936, pp. 419—424 (*4.34*, *7.26*).

FEJÉR, L., and RIESZ, F.: [1] Über einige funktionentheoretische Ungleichungen, Math. Z., vol. 11, 1921, pp. 305—314 (*7.25*).

FRAZER, H.: [1] Some inequalities concerning the integrals of positive subharmonic functions along curves of certain types, J. London Math. Soc., vol. 6, 1931, pp. 113—117 (*7.25*).

[2] An inequality concerning subharmonic functions in an infinite strip, J. London Math. Soc., vol. 7, 1932, pp. 214—218 (*7.25*).

[3] Further inequalities concerning subharmonic functions, J. London Math. Soc., vol. 7, pp. 284—290 (*7.25*).

[4] On the moduli of regular functions, Proc. London Math. Soc., vol. 36, 1934, pp. 532—546 (*7.25*).

FROSTMAN, O.: [1] Potentiel d'équilibre et capacité des ensembles, avec quelques applications à la théorie des fonctions, Thesis, Lund, 1935 (*6.21*).

GABRIEL, R. M.: [1] Some results concerning the integrals of moduli of regular functions along curves of certain types, Proc. London Math. Soc., vol. 28, 1926, pp. 121—127 (*7.25*).

[2] An inequality concerning the integrals of positive subharmonic functions along certain circles, J. London Math. Soc., vol. 5, 1930, pp. 129—131 (*7.25*).

GARRETT, G. A.: [1] Necessary and sufficient conditions for potentials of single and double layers, Amer. J. Math., vol. 58, 1936, pp. 95—129 (*7.22*).
HAHN, H.: [1] Theorie der reellen Funktionen, Berlin 1921, vol. 1 (*1.3*).
HAUSDORFF, F. [1]: Grundzüge der Mengenlehre, 1. Aufl. Leipzig 1914 (*4.6*).
HARTOGS, F.: [1] Zur Theorie der analytischen Funktionen mehrerer unabhängiger Veränderlichen, Math. Ann., vol. 62, 1906, pp. 1—88 (*3.37*).
KELLOGG, O. D.: [1] Foundations of potential theory, Berlin 1929 (*1.3, 1.6, 1.12, 1.14, 4.4, 4.25, 5.5, 5.6, 5.10, 5.11, 5.16, 5.17, 5.23, 5.24, 6.1, 6.21*).
KOZAKIEWICZ, W.: [1] Un théorème sur les opérations et son application à la théorie des laplaciens généralisés, C. R. Soc. Sci. Varsovie, vol. 26, 1933, pp. 1—7 (*3.7*).
KURATOWSKI, C.: [1] Une méthode d'élimination des nombres transfinis des raisonnements mathématiques, Fundam. Math., vol. 3, 1922, pp. 76—108 (*4.6*).
LITTLEWOOD, J. E.: [1] On the definition of a subharmonic function, J. London Math. Soc., vol. 2, 1927, pp. 189—192 (*2.3*).
[2] On functions subharmonic in a circle, part I, J. London Math. Soc., vol. 2, 1927, pp. 192—196 (*7.18, 7.21*).
[3] On functions subharmonic in a circle, part II, Proc. London Math. Soc., vol. 28, 1928, pp. 383—393 (*7.18, 7.22*).
[4] On functions subharmonic in a circle, part III, Proc. London Math. Soc., vol. 32, 1931, pp. 222—234 (*7.18, 7.23*).
MALCHAIR, H.: [1] Sur les fonctions sousharmoniques, Mém. Soc. Roy. Sci. Liège, vol. 20, 1935, pp. 1—15 (*3.5, 3.22*).
[2] Sur les fonctions sousharmoniques, C. R. Soc. Sci. Varsovie, vol. 28, 1936, pp. 71—76.
MONTEL, P.: [1] Sur les suites infinies des fonctions, Ann. École Norm., series 3, vol. 24, 1907, pp. 233—344.
[2] Sur les fonctions convexes et les fonctions sousharmoniques, J. Math. pures appl., series 9, vol. 7, 1928, pp. 29—60 (*2.8, 2.11, 2.17, 3.5, 3.12, 3.13, 3.17, 3.18, 3.20, 3.22, 3.37, 4.3*).
[3] Sur les fonctions doublement convexes et les fonctions doublement sousharmoniques, Prakt. Akad. Athenon, vol. 6, 1931, pp. 374—385 (*3.22*).
MORREY, C. B.: [1] A class of representations of manifolds, part I, Amer. J. Math., vol. 55, 1933, pp. 683—707 (*3.32*).
PÓLYA, G., and SZEGŐ, G.: [1] Aufgaben und Lehrsätze aus der Analysis, vol. I, Berlin 1925 (*3.13, 4.22*).
PRIVALOFF, I.: [1] Sur un problème limite des fonctions sousharmoniques, Rec. math. Soc. math. Moscou, vol. 41, 1934, pp. 3—9 (*7.18*).
[2] Sur un problème limite de la théorie des fonctions, Bull. Acad. Sci. URSS, series 7, number 2, 1935, pp. 301—304 (*7.18*).
[3] Sur certaines questions de la théorie des fonctions subharmoniques et des fonctions analytiques, Rec. math. Soc. math. Moscou, vol. 41, 1935, pp. 527—550 (*4.3, 7.24*).
[4] Sur la théorie générale des fonctions harmoniques et subharmoniques, Rec. math. Soc. math. Moscou, N. s. [1], 1936, pp. 103—120 (*2.14, 4.3, 7.24*).
RADÓ, T.: [1] Über eine nicht fortsetzbare Riemannsche Mannigfaltigkeit, Math. Z., vol. 20, 1925, pp. 1—6 (*7.26*).
[2] Remarque sur les fonctions sousharmoniques, C. R. Acad. Sci. Paris, vol. 186, 1928, pp. 346—348 (*3.12*).
[3] On convex functions, Trans. Amer. Math. Soc., vol. 37, 1935, pp. 266—285 (*3.27*).
[4] On the harmonic majorants of subharmonic functions, Bull. Amer. Math. Soc. vol. 42, 1936, p. 813 (*5.28, 5.33, 5.35*).

See also Beckenbach-Radó.

Radon, J.: [1] Theorie und Anwendung der absolut additiven Mengenfunktionen, Akad. Wiss. Wien, S.-B. Math. Nat. Kl., vol. 122, 1913, pp. 1295–1438 (*4.6–4.8, 4.11, 4.12, 4.15, 6.4*).

Riesz, F.: [1] Sur les valeurs moyennes du module des fonctions harmoniques et des fonctions analytiques, Acta Litt. Sci. Szeged, vol. 1, 1922, pp. 3–8 (*2.16, 3.20, 4.3*).

[2] Über die Randwerte einer analytischen Funktion, Math. Z., vol. 19, 1922, pp. 87–95 (*7.20*).

[3] Sur une inégalité de M. Littlewood dans la théorie des fonctions, Proc. London Math. Soc., vol. 23, 1924 (*4.3*).

[4] Über subharmonische Funktionen und ihre Rolle in der Funktionentheorie und in der Potentialtheorie, Acta Litt. Sci. Szeged, vol. 2, 1925, pp. 87–100 (*3.37, 4.3, 4.6*).

[5] Sur les fonctions subharmoniques et leur rapport à la théorie du potentiel, part I and II, Acta Math., vol. 48, 1926, pp. 329–343, and vol. 54, 1930, pp. 321–360 (*1.1, 1.6, 1.7, 1.9–1.13, 1.15, 2.4, 2.7, 2.9, 2.10, 2.16, 2.19, 2.24, 3.2–3.4, 3.6, 3.10, 4.3, 4.6, 4.18, 4.23, 4.26, 4.31, 4.34, 5.3–5.8, 5.10–5.14, 5.18, 5.20–5.22, 6.3, 6.4, 6.10, 6.15, 6.18, 6.19, 7.5*).

[6] Su alcune disuguaglianze, Boll. Un. mat. Ital., vol. 7, 1928, pp. 1–3 (*7.20*).

[7] Sur les valeurs moyennes des fonctions, J. London Math. Soc., vol. 5, 1930, pp. 120–121 (*2.12*).

[8] Eine Ungleichung für harmonische Funktionen, Mh. Math. Physik, vol. 43, 1936, pp. 401–406 (*7.25*).

See also Fejér-Riesz.

Saks, S.: [1] Sur une inégalité de la théorie des fonctions, Acta Litt. Sci. Szeged, vol. 4, 1924, pp. 51–55 (*7.26*).

[2] Sur un théorème de M. Montel, C. R. Acad. Sci. Paris, vol. 187, 1928, pp. 276–277 (*3.17*).

[3] On subharmonic functions, Acta Litt. Sci. Szeged, vol. 5, 1932, pp. 187 to 193 (*3.7, 3.15, 3.16*).

[4] Note on defining properties of harmonic functions, Bull. Amer. Math. Soc., vol. 38, 1932, pp. 380–382.

[5] Théorie de l'intégrale, Warszawa 1933 (*1.4, 3.7*).

Szpilrajn, E.: [1] Remarques sur les fonctions sousharmoniques, Ann. of Math., vol. 34, 1933, pp. 588–594 (*3.7, 3.28, 3.33, 3.34, 3.35, 3.36*).

Thompson, J. M.: [1] Distribution of mass for averages of Newtonian potential functions, Bull. Amer. Math. Soc., vol. 41, 1935, pp. 744–752 (*6.23*).

Valiron, G.: [1] Remarques sur certaines fonctions convexes, Proc. Phys.-Math. Soc. Jap., series 3, vol. 13, 1931, pp. 19–38 (*3.22*).

Wiener, N.: [1] Laplacians and continuous linear functionals, Acta Litt. Sci. Szeged, vol. 3, 1926, pp. 7–16 (*6.20*).